GERONTECHNOLOGY

GERONTECHNOLOGY

Understanding Older Adult Information and Communication Technology Use

BY

JOHANNA L. H. BIRKLAND
Bridgewater College, USA

United Kingdom – North America – Japan – India – Malaysia – China

Emerald Publishing Limited
Howard House, Wagon Lane, Bingley BD16 1WA, UK

First edition 2019

British Library Cataloguing in Publication Data
A catalogue record for this book is available from the British Library

ISBN: 978-1-78743-292-5 (Print)
ISBN: 978-1-78743-291-8 (Online)
ISBN: 978-1-78743-949-8 (EPub)

An electronic version of this book is freely available, thanks to the support of libraries working with Knowledge Unlatched. KU is a collaborative initiative designed to make high quality books Open Access for the public good. More information about the initiative and links to the Open Access version can be found at www.knowledgeunlatched.org

Printed and bound by CPI Group (UK) Ltd, Croydon, CR0 4YY

ISOQAR certified Management System, awarded to Emerald for adherence to Environmental standard ISO 14001:2004.

Certificate Number 1985
ISO 14001

To my children: for reminding me the joys of curiosity and always giving me a reason to adventure.

To my husband: for being an equal partner (in crime).

Contents

List of Figures

List of Tables

About the Author

Johanna L. H. Birkland is an Assistant Professor in Communication Studies at Bridgewater College in Bridgewater, Virginia, USA. An interdisciplinary Gerontechnology scholar, she uses a mixture of approaches to study how technology use changes over the life course as well as intergenerational technology use and communication.

Acknowledgments

This research would not be possible without the participants who invited me into their homes and lives: thank you for sharing your stories with me. I appreciate the support of my current Bridgewater College colleagues, too numerous to list. Many thanks to Sarah Chauncey and Fatima Espinosa Vasquez for their numerous discussions about the material within these pages, and for being friends in addition to colleagues. I want to recognize and thank my mentors and those who guided my development as a researcher, most notably Michelle Kaarst-Brown, Janet Wilmoth, Renee Hill, and Steve Sawyer. I gratefully acknowledge the financial support of the Bridgewater College Faculty Research Fund and the Syracuse University School of Information Studies Katzer Doctoral Research Fund. Thank you to the support team at Emerald, most notably Jen McCall and Rachel Ward. On a personal note, I want to thank the late Kathy Berggren for her support and mentorship. Finally, I owe my husband, Aaron, and my children, a debt of gratitude for supporting me, always.

Chapter 1

Understanding Older Adult Technology Use: An Introduction to the ICT User Typology

Gwen,[1] born in 1946, is a retired administrative assistant who wants all the technologies that "young people have." She has a large family, with five children, over 20 grandchildren, and several great-grandchildren. Highly involved in her community, she volunteers 20 hours a week as a teacher's aide in an inner-city school, is active in her multi-generational church, and runs a small food pantry for her neighbors out of her apartment closet. Many of the neighborhood children in her low-income housing complex call her "Grandma Gwen." Gwen is very interested in the technologies that young people use to communicate, including cell phones, social media, and texting. She is a prolific user of these communication technologies and mimics the use of the young people she knows:

> I am the queen of texting. I have to text. I'm forced to do texting because some of my grandchildren just will not answer the telephone. They have their phones on vibrate so they just will not talk on the phone. So, if I want to ever talk to them I have to text. I'm a great texter and I know all the abbreviations. I make some of them up myself and I have them ask me what they mean [...] I don't know what they say or think about me doing all this texting, but I love to do it. And I have to do it. (Gwen)

Margaret, born in 1938, has recently retired from her work as an administrative assistant, a role she returned to in the 1980s when she divorced after 20 years of staying home with her children. She enjoys gardening and attending her book club and has just volunteered for a political campaign. Margaret carefully regulates her use of the computer, the Internet, and her simple flip phone, often setting a timer to limit how long she stays on her computer. She fears that without such restrictions, she could be "sucked in" – losing track of time and disengaging from the world around her – a potential problem she sees with all technology use, including the television and telephone. She keeps her computer

[1]All participants' names are pseudonyms.

in the basement den and a television in the smallest spare bedroom to keep her main living space technology-free. She is cautious about technology and feels that it is often overused:

> I feel that there's a need, there's definite need for all this modern technology. There's a need for it but I think it's just like many, many things it's overdone. I think it's absolutely mind-boggling ridiculous that cars now have TVs in them. Look out the window. Enjoy, see what you're seeing. It's removing them [technology users] from a part of life that I think is important [...] you go to the mall and you see people walking around and they're just talking on the cell phones. Talk, talk, talk on the cell phone. I thought you went to the mall to go shopping. So, I think that people go overboard on all that stuff. (Margaret)

Why are these two women so different? Why has Gwen embraced technology with such enthusiasm, but Margaret is cautious and controls her use of computers and cell phones? How do these women's different approaches toward technology impact their lives, and how did the meanings they hold for Information and Communication Technologies (ICTs) develop?

This book presents a theory of ICT use by older adults, the ICT User Typology, which not only explains the diversity in older adult ICT use, but also helps practitioners, scholars, and designers to understand the older adult population's needs and wants when it comes to technological interactions. This data-driven theory emerged from the findings of a rigorous, in-depth interpretative interactionist (Denzin, 2001) series of comparative case studies (Yin, 2009) of the use of ICTs by community-dwelling older adults (ages 65–75) in their everyday lives. The ICT User Typology describes five ICT user types – each of which has a unique view on technology and uses it in different ways. The focus of this book is in detailing these five types and understanding their fundamental traits. While the ICT User Typology is a data-driven theory, it has many practical applications for those who work with or design for older adults, including suggesting targeted design and marketing opportunities and identifying those older adults who are likely to take part in (or be excluded by) technological initiatives. (Practical applications of the theory are outlined in Chapter 10.)

The ICT User Typology: The Five User Types

The ICT User Typology is a theory of older adult ICT use that explains, describes, and predicts an older adult's use of multiple ICT forms across various life contexts: family, work, leisure, and community. The ICT User Typology was developed from an intensive series of dialogic case studies of 17 older adult members of the Lucky Few generation, including over 156 hours of intensive interviews with these older adults and their friends, family members, and coworkers about ICT use; numerous observations of their homes and workplaces and of the older adults using technologies; and careful review of the documents that the older adults used in conjunction with their ICTs. The typology allows us to understand how and why there is such diversity in older adults' ICT use and provides guidance in implementing customized services and products to meet the needs of our aging societies. In particular, the ICT User Typology categorizes older adults' ICT use into one of five user types, each of which has a unique pattern of ICT introduction, use, display, and meaning they ascribe to technologies:

The *Enthusiast* user type thinks ICTs and other forms of technology are great fun toys. They have wonderful memories of using ICTs as children, including being encouraged by adults to "tinker" and "play" with technology. They carry this love of ICTs throughout their lives, often choosing technical careers. They surround themselves with other Enthusiasts as friends and, in some cases, become romantically involved with other Enthusiasts.

The *Practicalist* user type views ICTs as tools that are used to get a job done, for a specific purpose. They are typically exposed to ICTs in their work and they tend to hold jobs in which technology is heavily used. They are easily able to categorize their ICT device use into predominantly being used for one life context, such as being able to state that they use the Internet mostly for work, rather than in their leisure, family, or community lives.

The *Socializer* user type tends to have large intergenerational networks and be highly involved in their communities, often through religious organizations and/or large families. They view ICTs as connectors between people and tend to prefer mobile communication technology. Socializers, in order to keep in touch with their youngest contacts, learn to use the ICTs that their younger counterparts are using.

The *Traditionalist* user type also speaks about their love for ICTs. However, the technologies that Traditionalists love are the ones from their young adulthood (in the case of the older adults spoken about in this book, the television, radio, and telephone). They have a tendency toward nostalgia and find themselves so in love with these more traditional technologies that they have little room in their lives for more modern forms (e.g., computers and cell phones).

The *Guardian* user type tends to view *all* ICTs with suspicion, as they believe that technology can bring out the negative traits in individuals – traits such as gluttony and laziness. While they use many modern forms of advanced ICTs, they tend to be very cautious and regulated in how and how much they use them. They view themselves as protectors, or guardians, of society in its use of ICTs.

Why Is a Theory of Older Adult ICT Use Necessary?

Why do we need to understand older adult ICT use? The short answer is that the vast majority of our societies are aging. The worldwide older adult population (those age 65 and older) is projected to nearly double in the first third of this new century – from only 6.9% in 2000 to 12% in 2030. The more detailed figures are striking: the population percentage of older adults is expected to increase from 15.5% to 24.3% in Europe, 12.6% to 20.3% in North America, 5.5% to 11.6% in Latin America and the Caribbean, and 6.0% to 12.0% in Asia during those same three decades. Even in sub-Saharan Africa, where high fertility rates have continually bolstered the younger population, the percentage of older adults is still expected to increase from 2.9% in 2000 to 3.7% in 2030 (Kinsella & Velkoff, 2001).

At the same time, our aging societies are facing an increasingly digitalized world. Services and organizations, including governments, have increasingly moved information and resources online – and in some cases, these can only be accessed digitally (Wright & Hill, 2009). Older adults (those aged 65 and older) report significantly lower usage of advanced ICTs such as computers and smartphones when compared to younger individuals (Heinz et al., 2013; Reisenwitz, Iyer, Kuhlmeier, & Eastman, 2007; van Deursen & Helsper, 2015; Van Volkom, Stapley, & Amaturo, 2014). These two phenomena taken together – lower use rates and increasing digitalization – have led to concerns that a "gray digital divide" will exclude elders (Friemel, 2016; Millward, 2003) and has been portrayed as part of the larger "aging crisis" facing our societies (Wright & Hill, 2009).

But do our aging societies represent a crisis, or an opportunity? Are older adults able to get access to the digital products and services they need, and what about the products and services they want? How can organizations take advantage of this growing market? How can we understand older adult technology use, and how can we understand the diversity in such use? What *are* the characteristics of the older adult market for ICT[2] services and devices?

The power of the ICT User Typology is that it allows classification of older adults into one of these five user types, helping us to understand how each of these types is introduced to, uses, and finds meaning in technology. For researchers, the ICT User Typology presents a theory of ICT adoption and use that may be insightful to apply in their own studies. For practitioners who work with older adults, the ICT User Typology presents an overview of the different

---

[2]The ICT User Typology was developed based upon the study of ICTs that are marketed to the general public. This includes both physical ICTs (e.g., cell phones and computers) and software and applications (e.g., social networking applications and computer programs). ICTs include technologies such as computers, telephones (landlines), cell phones, Internet, email, social networking software and applications, radio, television, and print media (magazines, catalogs, and newspapers). Internet and computer applications are also considered ICTs, such as Skype, word processing, digital image software, and so on. The theory also encapsulates assistive technologies that allow individuals to access the information or communication potential of these devices or media, such as "TV Ears" (a device that allows a hard of hearing listener to listen to a television). Assistive devices that do not allow a person to access a communication or information device (such as ramps, wheelchairs, and oxygen tanks) are not included.

ways older adults approach ICTs in their daily lives, illustrating which older adults are likely to take advantage of digital products and services and which older adults are at risk of exclusion, particularly if such use is required. For designers, the ICT User Typology guides the targeted and customized development of ICT products and services for the older adult market.

The ICT User Typology: Use as Domestication Patterns

The ICT User Typology suggests that there are five unique and easily identifiable domestication patterns among older adults. Domestication is the process by which ICTs are brought from the larger outer world into individuals' lives, developing routines of use and meaning (Silverstone & Hirsch, 1992; Silverstone, Hirsch, & Morley, 1994). Domestication looks beyond the simple adoption or ownership of a device to understand ICT use in a context-rich environment (Haddon, 2007; Lie & Sørensen, 1996; Silverstone & Haddon, 1996; Silverstone et al., 1994) as a series of steps:

- ICTs are first *introduced* to, or "appropriated" into a setting. During this time, ICTs are removed from the "public sphere" to acquire meaning in the "private sphere" of the home. Meanings in the private and public spheres are not necessarily similar: The same ICT can have vastly different meanings to different people.
- The ICT is *displayed* and arranged in the home, and through the process of objectification, certain individuals in the home come to identify with and feel comfortable with these technologies, while others do not.
- The ICT is *used (or not)*, and this use is impacted by larger contextual factors in the home.
- The ICT and its placement in the home (or on the person) develop *meaning* to both the individual and the larger social society and help in both self-identification and the identification of others, as users and non-users.

As a broad theory of ICT domestication, the ICT User Typology demonstrates the importance of taking a multiple ICT perspective to understand the older adult use and non-use. While a few studies have examined how older adults use multiple ICTs (Hilt & Lipschultz, 2004), the existing literature has typically focused on exploring use and non-use of a single ICT per study (typically computers and the Internet (van Deursen & Helsper, 2015; Xie & Jaeger, 2008)). Such a perspective does not illustrate how a non-user compensates for non-use or how older adults integrate the use of various ICTs together. Instead, The ICT User Typology focuses on how older adults incorporate, reject, and integrate the spectrum of ICTs into their lives. For instance, the Traditionalist user type in older age relies on one or two "point people" to access the more modern ICTs they do not use, while they deeply appreciate and heavily use the more traditional ICTs of their youth (radio, television, and telephone).

The ICT User Typology: A Context-rich Gerontechnological Theory

The ICT User Typology explores ICT use across older adults' entire life contexts: their family, work, community, and leisure lives. Most importantly, it captures and describes older adults' perspectives on their use, what makes them want to (or not want to) use a device, and what motivates them and why in their own words.

Gerontechnology, or the interdisciplinary study of gerontology and technology, has existed since 1989 (Graafmans & Brouwers, 1989). Gerontechnologists come from a diversity of fields, ranging from the sciences to social sciences; and span academics, clinicians, and other practitioners. As can be imagined, the literature in this area is quite diverse, being influenced both by the background of the researcher(s) and by the vast number of questions that can be asked about aging and technology.

The goals of Gerontechnology research and application include not only preventing, accommodating for, and delaying the cognitive and physical declines related to aging, but also for using technology to enhance the lives of older adults. Beyond the fundamentals of being well-cared for, Gerontechnology also includes studies of how technology can be used to promote older adults' life satisfaction, communication, social connectivity, and volunteerism (Fozard, Rietsema, Bourma, & Graafmans, 2000). Gerontechnology has tackled using technological solutions to provide health care to our aging population (Czaja, 2016; Fischer, David, Crotty, Dierks, & Safran, 2014), smart home solutions (Majumder et al., 2017), and general guidelines for creating adaptive and inclusive technologies (Bouma, 2001). Gerontechnology has called for researchers to study all the areas of older adults' lives: their health, family, community, leisure, and work lives (van Bronswijk, Bouma, & Fozard, 2002; van Bronswijk et al., 2009). Recently, there have been calls for Gerontechnologists to not just understand what ICTs older adults are adopting, but how and why they are using them (Schulz et al., 2015). The ICT User Typology addresses both the *how* and the *why* of older adult ICT use.

One important contribution that the field of Gerontology has made to the study of Gerontechnology has been the importance of understanding ICT use in terms of generations. Birth cohorts are groups of individuals who, by consequence of being born closely together, experience historical events at a similar life stage, developing a shared generational consciousness (Carlson, 2009; Edmunds & Turner, 2002; Eyerman & Turner, 1998). Media and technology is an important part of any generation's experience (Naab & Schwarzenegger, 2017). The ICT User Typology was developed using generational sensitive sampling of the youngest generation of individuals who had reached older adulthood in the United States: the Lucky Few (participants ranged in age from 66 to 76 at the time of the study (born in 1936–1946)). The Lucky Few are a small generation born during the Great Depression and World War II(WWII). They tended to serve in the military during peacetime, while enjoying the benefits created for the WWII Generation that served before them, such as the GI Educational Bill.

The Lucky Few women made tremendous strides in joining the workforce compared to previous generations; however, this was mainly in "pink-collared" professions (as nurses, administrative assistants, teachers, etc.) (Carlson, 2008).

Although the ICT User Typology was developed using a single generation of older adults, it is applicable to all generations/birth cohorts. Evidence from interviews with the older adults' families, friends, and coworkers indicates that these five user types are universal and likely found in all generations at mid-adulthood and beyond. These additional individuals interviewed in this study ranged in age from 27 to 85 (Millennials to the WWII Generation), and their perspectives on ICT use echoed the five user types described in this book. It is likely that a person's user type is influenced by childhood and early adulthood interactions with technology, regardless of a person's generation/birth cohort.

While technological advances do not influence the essence of the types (Guardian*s* will always be suspicious of all technology; Enthusiasts will always want the latest and greatest devices), the technologies an individual of a specific type prefers will reference their generational technological experiences. For instance, Traditionalists prefer the technologies of their youth. For members of the Lucky Few birth cohort (born in 1929–1946), the technologies of their youth were radio, television, and telephones. For members of the Millennial birth cohort (born in 1983–2001), the technologies of their youth were computers, cell phones (including smartphones), and social media (in addition to television and radio). Therefore, a Millennial Traditionalist will prefer the media of his/her youth (cell phones, social media, etc.) rather than newer technologies that develop in the future. What makes a Traditionalist is *not* their non-use of computers, but instead, the fact that they reject forms of technology that were not available in their youth. Chapter 8 explores the implications of the user typology for younger generations in-depth. Since the influences of childhood and early adulthood are incredibly important in shaping individual's tendencies to become one type over another, Chapter 10 details the implications of the ICT User Typology for both gerontologists and childhood educational programs.

Chapters 2 through 6 deal with an extensive description of each of the five user types: the Enthusiast, Practicalist, Socializer, Traditionalist, and the Guardian. In these chapters, we hear directly from older adults. We peek into their homes and workspaces and come to understand ICT use from their own perspectives. Chapter 7 summarizes and contrasts the five user types. Chapter 8 explores the applicability of the ICT User Typology beyond the Lucky Few generation, exploring its impacts across generations. Chapter 9 grounds the ICT User Typology within other theoretical perspectives, understanding its fit within the gerontechnological and technology use literature. Chapter 10 provides recommendations for applying the user typology for practitioners and scholars. Chapter 11 provides a detailed methodological description of how the ICT User Typology was developed, allowing further study and possible replication not only in older adult populations, but also in younger populations as well. A Glossary of terms is included.

The next chapter takes us into the lives of Enthusiasts to understand their love of "all things with plugs" (Alice).

Chapter 2

Enthusiasts: The Technological Evangelists

> Oh, I love technology. I have ever since I started using it way, way back when. But I just fell in love. I love everything from the word processor to the projector to making film strips. Recordkeeping is so easy and when I discovered spreadsheets, I was just in love. I've enjoyed the advantages of this kind of thing ever since they started making it available. I'm like a little kid in a candy store. I love to play around with everything – I just love this stuff. Love it. Love it! (Fred)

Love. Play. Fun. Toys.

To hear an Enthusiast speak about Information and Communication Technology (ICT) is to listen to a love ballad and, with the other ear, to listen to a five-year-old bursting with excitement in a candy store. Fred, in his edited quote, represents the essential Enthusiast relationship with technology: one of adoration and excitement. In the original transcript, Fred continues for two pages to talk about all the technologies he loves. When you speak with an Enthusiast about technology, you open the floodgates to hearing about their passion – and their passion is *most definitely* technology.

Enthusiasts' lives center on ICTs. They tend to have been exposed to ICTs early in childhood, in mostly positive interactions. They push technologies in their everyday lives to be used across work, family, community, and leisure tasks and relationships. They tend to form close bonds with other Enthusiasts, while filling their homes with ICTs in prominent places. Above all, ICTs invoke feelings of excitement – and using an ICT is much more play than work.

Formative Experiences

Enthusiasts tend to point toward a lifelong interest in technology that started at an early age, oftentimes with encouragement from their family members and older friends to explore technologies. Enthusiasts of this generation were the kids who often took apart their televisions, created kit radios, and modified their cars. When asked about their relationship with ICTs, they often share vivid memories from childhood about encountering a new technology for the first time. For instance, Fred spoke about his first encounter with television:

> I remember the first TV I saw [...] it was a little TV screen, in a big box. It was black and white. In the Fifties there were only three channels in my city. After 11:30 at night the only thing on was a test pattern. That was the early Fifties. It was amazing and so nobody on our street had a TV. And then the one kid I hung around with on the street their family got a TV. We'd go down there and watch TV and they had the fights on Friday and that was about it. But it was amazing. Later they had movies on TV, you could watch the news [...] it was great! (Fred)

While Fred speaks about how limited early television was (with only the fights broadcast on Fridays), he also speaks in terms of amazement and enthusiasm. In their descriptions, Enthusiasts focus on the technology itself and their personal relationship with it. While other types see technology as being a tool to get something done or a connector between people, Enthusiasts think using technology (even for mundane tasks) is simply fun play. In childhood, this fascination was often encouraged by those around the Enthusiast. In particular, Enthusiasts were encouraged to "tinker," to take technologies apart and put them back together:

> My father was a mathematician engineer and he was career Army. He was into technology and electronics, so he would come home and bring radios and all kinds of stuff. So, I guess from the time I was a kid there was stuff around to play with. He would let me play with anything he brought home, and I just got into it. I've always just had this thing for playing with technology. (Fred)

Fred speaks about how his father was instrumental in his early love of technology by encouraging him to "play" with new gadgets. Similarly, Harry speaks about his grandfather, who tended to be on the leading edge of technologies and who let Harry play with his shortwave technology and other things that were available in his service department:

> My grandfather ran the service department at a car dealership in the 1920s. My grandfather was an early adopter of technology and I grew up and around his shop. So, I think I grew up with the love of technology. My grandfather was always messing with new things. There was a shortwave radio that was my granddad's and it was a Zenith which in the 1950s was an equivalent of the iPod. I always used it in his office and when he passed away I got the radio. So, I grew up in a technical environment, around cars being fixed and electronics. I was interested in electronics when

> I was a kid. My grandfather was the first kind of direct influence, and I don't really think anybody in particular really matched his influence on me. I got into ham radio in the Boy Scouts and I built a ham radio from a kit and did ham radio stuff and I really think that is because of my grandfather's encouragement. (Harry)

Enthusiasts, unlike other types, tend to be able to point to one person who has been pivotal in their journey with technology, resulting in the development of what has become a lifelong love affair with "things with plugs" (Alice). Harry could point to his grandfather as being the most influential person in his relationship with technology, while Fred pointed toward his father. At times, these relationships could be quite fragile due to other factors, but the love of technology helped to maintain a connection. Fred shared that his relationship with his father was strained as Fred had chosen the career path of a teacher and an artist (and later an IT professional) rather than an engineer, as his father had wanted. However, throughout their relationship, they could always find a common ground over technology.

As Enthusiasts grow older, their "toys" tend to become bigger. Male Enthusiasts often turned toward "hot rodding" their cars by adding speakers and trying to make them faster:

> I was also fiddling in high school and college with car radios trying to beef up my car radio to make it not one speaker like they all had. I wanted five speakers if I could put them in. I've always just had a thing about machinery. (Fred)

> So, when I got out of the service I ended up going to college and I needed money. I gravitated towards cars, so I started selling cars for this little sports car dealership in a Midwest City. Some of the guys that were racing cars were having trouble keeping their cars running so I started making deals with them that I would rebuild their engine in return for being able to use their car to go through driver's school. You had to go through four or six races in driver's school and then you had to do six novice races. So, the deal was that I would fix their car and for racing weekend I would use it for the novice race and they would use it for the more advanced race, so it worked out pretty well. So, I ended up starting racing cars by working on them, I fixed them and stuff. Then after college I won a major motor race and I went off on kind of this quixotic exercise of becoming a professional racecar driver which didn't work out. But in the course of it I got quite a reputation for fixing racing cars and then street cars, so I ended up with a car shop in a [North Eastern State]. I became the go to guy for exotic cars and stuff like that because

> they required a little higher level of mechanical insight to do that… so I've always been interested in technology. Mechanics and technology. (Harry)

As Enthusiasts grow older, their interest in technology leads them to careers that often have a heavy technological focus. Enthusiasts see technology as being an important part of their career trajectory and often credit their love of technology as being the determining factor in their career choices. Harry, for instance, credits technology with "*saving*" him from becoming a delinquent, leading him instead on a career path to becoming an Information Technology (IT) professional:

> I really thought that when I grew up I wanted to be an aeronautical engineer. I didn't have a real drive for it and so in a way technology saved me […] I was a very bright student but I did not do well with authority. I had a lot of problems at home and I wasn't quite a delinquent, but I probably wasn't far from becoming one. The reason I ended up in the Army was I wanted to be a pilot and no other service would take someone without a high school diploma. The Army said, "since you're a little young for flight school you can go to helicopter mechanics school first." So, I became a helicopter mechanic and in that time I ended up not going to flight school. I ended up realizing three years in the Army was going to be enough. So, I had a very vocational focus but a pretty high level education in both aircraft mechanics, structures, and in electronics for doing the systems and stuff like that. Then I went on to fix cars, because of my electronics and mechanical background […]
>
> [Later] I started working in the physics lab. I'd always been interested in electronics and stuff. I built radios as a kid and I had gotten the electronics experience in the military. One of the reasons I got the job in the physics lab was being able to do electronics troubleshooting and maintenance; design and build power supplies, and stuff like that. That put me in touch with working on systems and that's how I came to do IT systems work. (Harry)

While not all Enthusiasts become IT professionals, all the older adult IT professionals in this study were Enthusiasts, and all Enthusiasts choose to incorporate heavy use of ICTs into their work lives. Alice, the only Lucky Few female Enthusiast in the study, had a slightly different journey with technology throughout her life. Alice was encouraged to be involved with technology from a young age, and in adulthood she chose to become a nurse, at one time working as a medical administrative assistant. While most would not see technology taking a central point in this type of work, Alice noted that she often "pushed" technology in her workplace to be as efficient and effective as possible, even early in her career:

> Many years ago, I was a nurse secretary in a big office. I had an electronic typewriter, which I programmed as much as I could. It made my job so much easier and efficient. When I left that office, there was one secretary who was really a bitch and she said "oh when you leave I'd love to have your computer – your typewriter!" I said, "oh sure!" I brought it into her and I said, "Here's my typewriter," and what she didn't know is that I had erased all the programs I had put on there. I knew that she couldn't figure that out. I've always pushed the technology to make it work for me. (Alice)

While most of the administrative assistants Alice worked with did not spend time programming their typewriters, she spent extensive time researching and finding out how she could program various shortcuts into the system. Alice greatly enjoyed this task, as well as how it benefited her during her work.

Introduction to ICTs

Enthusiasts are the most likely, of all the types, to seek out new ICTs (or to update their ICT forms to the latest version) on their own, often out of curiosity. Enthusiasts rely heavily on media to stay aware of new ICTs and updates, as well as to learn new techniques and methods for using their ICTs. In particular, they rely on technical columns, blogs, and magazines:

> I had been reading *Byte Magazine* for years. I'd say Byte was a big mentor for me, that's how I got started in technology when I came to the area [...] I still read a lot of technical blogs. (Harry)

> I read about stuff. I get the local paper so I read all the tech columns when they come out. Then often Fred and I discuss what I read. (Alice)

Alice speaks about the importance of not only reading about new ICTs, but also sharing that knowledge with others. Enthusiasts are the most likely of all the types to try out something new, to play with a new technology, and experiment with it:

> I'm bolder than Fred is so I will go out and try something before he does, or I'll put my foot down and say "no, I want to do this." He's always gifting me technology. Quite a few years ago my small color photo printer died, and he said, "well I'll get you another one" and he's looking at mid-priced ones. I had been

> working with another photographer and I learned about the more expensive printers. So, I told Fred I'm going to buy the printer because I didn't want to ask him to buy a very expensive one. About two years after that they had come out with an even better one so then he bought that. It was the same thing with the Smartphone. Once we have the same stuff, we learn from each other. (Alice about Fred)

Fred and Alice, Enthusiasts who were life partners (and both participants in this study), demonstrate an important aspect of technology to Enthusiasts: it underlies many of their relationships and their shared interest in technology helps them to form a bond of friendship, and in some cases (such as Fred and Alice), even romantic partnership. Enthusiasts share technologies and learn from one another. Unlike the other types, however, the introduction to ICTs in these relationships is not one-sided. Relationships with Enthusiasts evoke technology sharing between both individuals, and in some cases, this technology sharing is the *basis of the relationship*. For instance, Fred shares how Tom first introduced him to the Internet in the early 1990s:

> Well Tom and I were friends to begin with, so it was a mutual interest in technology [...] It was shortly after we met that we both had 286 computers. Tom started on Prodigy [an early Internet service] before I did, actually he was the one that talked me into giving it a shot. Well I went to his house and I had my original computer with two big 5¼ inch floppy disks. I'd see Prodigy advertised and I thought "well that's sort of interesting but I don't know." So, I'd go to Tom's house and he'd go and show me Prodigy. He had email which was something new. There were groups that you could join, so if you're a photographer you could get [in a group] with all the photographers. After he showed it to me a couple of times and I thought, "Gee it's only $4 a month I can do that" and so I got into it. Then it was a case of every five months getting a new modem that was just a tiny bit faster. So, we were both interested in the computer. I guess he has an affinity for machines too. Actually, we started with our shared interest in photography, but our love of technology grew from there. (Fred about his friend Tom)

Enthusiasts are eager to hear how their friends are using a technology and encourage their friends to try their latest technological discoveries. Tom showed Fred the Internet every time Fred came to his house, encouraging him to try it, since it was low risk. Friendships (and relationships in general) tend to be an important part of how Enthusiasts are introduced to new technologies. In fact, Enthusiasts tend to prefer to have many of their relationships with others who are technically savvy. In some cases, technology can become the basis for romantic relationships. Fred and Alice met each other online, in the early days of Internet message boards:

> In the old days of Prodigy there was a photography board that I was on and this lady comes on looking for information about venting a dark room. So, I sent her some information. So, we message back and forth a little bit, and then maybe a year later or quite a few months later here's this person looking for post-mortem photography. I've been into cemetery things forever too, so I sent her a couple of emails on where to find some good cemetery stones and I suggested "well, why don't we get together?" We decided to meet because we lived close to each other. I said "well, would you like me to take you to a cemetery?" I know how to show a girl a good time. (Laughing.) So, we went that day and it just started a friendship. This was 15 or 16 years ago. (Fred on his relationship with Alice)

> I met Fred online, on a bulletin board. First, I was asking for help with my darkroom. Then later I was doing a class at my local community college on death and dying. I was interested in the use of photography in death and mourning. So, I went back to the bulletin boards and the same man who helped me with the dark room said "I don't know if you remember me, but I helped you with your darkroom a while ago. I know some neat places in nearby cemeteries." So, we met in person and we were friends first. Then we fell in love. (Alice on her relationship with Fred)

ICTs, their use, and sharing technology remain an important part of Alice and Fred's relationship to this day. Enthusiasts enjoy receiving and giving technological gifts; to Enthusiasts, the best type of gift is a technology:

> The last thing Fred got me as a gift was the new laptop. My bank is very good at Internet security. A service representative there called me and said "we were tracking your purchases and we had one that does not sound right. It was yesterday morning and it was a $200 withdrawal from an ATM on Main Street." And I was thinking where is Main Street? Then I realized it could be Fred! I asked the service representative for the account that it was drawn on. It was my joint account with Fred and I said, "no that's fine." So then when I called Fred about it he said some bad words and he said, "they've ruined my surprise!" The next day I come downstairs and there's the laptop with gift bows stuck on it! (Alice)

Alice and Fred spoke about many of the gifts they had bought each other over the years. As Enthusiasts, their most memorable gifts to one another were "things with plugs" (Alice), and included smartphones, laptops, printers, and cameras. In many ways, gifting these ICTs is important not only on the gift giving occasion, but because Fred and Alice spend a significant portion of time using, discussing, sharing, and teaching each other new things about technology; these gifts strengthen the relationship. In many ways, technology gifts represent a physical manifestation of their shared love of technology and its importance in their relationship.

ICT Use

Enthusiasts, noted by their love of ICTs, are interested in finding ways to use all of their ICTs in as many ways as possible, but also in moving as many processes from paper-based systems to digitalized ways of completing the same tasks. For instance, Alice speaks about how she moved her recipe collection to a digital format:

> I started about six years ago putting as many recipes as I could put on my main computer. And I printed out a loose-leaf notebook for my daughter. Then I had a law student living here with me for a while and when she graduated I gave a similar one to her. I got to thinking, "why am I always printing out all this stuff? I have a computer." Fred gave me the little computer, a notebook I guess it's called, so I brought all those digital recipes down and put them on the notebook. So now all I do is if I'm looking for something that I don't have a recipe for, I just take the notebook and I go on the Internet. Then I cut and paste what I find into a Word document, and then I've got it right there. So, I don't even really bother printing anything out anymore. (Alice)

Enthusiasts view using ICTs as fun and play, so their perspective is: why not have more fun and play in every aspect of my daily life? To achieve this, they are constantly investigating and experimenting with ways to integrate a single ICT across all the different areas of their lives: family, community, work, and leisure. Like Alice with her recipes on her notebook computer, Enthusiasts constantly look at ways to update and improve their lives through technology.

Alice originally had purchased her smartphone for personal use (inspired by Fred's purchase). She found many uses for her smartphone in her family and leisure life, but soon also discovered uses in her work as a home healthcare nurse:

> My smartphone is amazing. I'm always using it for the Internet. It's got really neat games! I'm a game person. But I use it in other areas of my life too. I was doing a temporary nursing case with a woman. Among other things I had to take her vital signs every day. I got there one day and darn it if my watch battery hadn't stopped. I just very quickly got on the smartphone, downloaded an app for an analog watch, turned it on, and there I was able to

> take the vitals and I was done! But I use it in other ways too. I started taking a couple pictures and showing them to one young patient and the next thing I knew she was picking up the phone on her own and looking for them. I'm always taking pictures and printing them out for her. I take the pictures on my phone because she can use my iPhone. She's got the mentality of a pre-kindergartner or maybe 1st grader but she uses the iPhone [...] it's been a great use of my iPhone. So, I take pictures of situations that make her uncomfortable. For example, she does not like the dentist. Her father has this habit of not telling her what's going on just saying "get in the car we're going for a ride," and it could turn out to be something fun like going to the mall or could be going to the dentist. I don't think that's fair, so I will tell her in advance this is what we're going to do. So, if I have pictures of it it's much better, so the last time at the dentist I took pictures of her in the dental chair and they gave her a latex glove that she loves. So now she goes to those pictures all the time and she looks at them and says, "Glove, Dr. Smith." So that's a good prep for her, I do that on a lot of different things with her. (Alice)

Enthusiasts are the most likely of any of the types to be serious in playing digital or virtual games: all of the Enthusiasts in the study stated that they played digital games of some type. Alice was the most serious gamer and, through online gaming, had made several enduring friendships. While Alice began using her smartphone mostly for games (leisure activities), she quickly found uses for it in her work life. At first, these were relatively simple functions (such as replacing her watch) but eventually grew to her using her phone in more complex situations, such as a visual aid for her home healthcare patient.

Enthusiasts' use of ICTs is not simply spread across various areas of their lives, but they are constantly looking for ways to "stretch" the use of a single device to different life contexts. Like Alice looking for ways to use her smartphone in her work or her notebook computer in cooking, Harry and Fred also looked for innovative ways to "stretch" devices in their daily lives:

> I use my computer all the time. I now have four computers, I have my base computer which is my big desktop and then I have two laptops, three laptops. Well two laptops and a Smartphone, which I consider a computer. I use them to do things. I do an awful lot of image processing, web searching, research, etc., etc. I use them for teaching and presenting, they're just a part and parcel. I have a projector when I'm talking somewhere that does not provide a projector. I use my computers to produce things for teaching classes. I use my computer for producing things, for putting talks together, teaching, for making little teaching units. I am just now dipping my toe into the not

> podcasting but learning how to actually put good audio on my computer via microphone and using a little mixer and whatever. I use it for entertainment; a lot of what I look at on my computer is strictly entertainment. I start my day by reading Arts and Letters Daily, Salon, Slate, New York Times, the Wall Street Journal, just to get myself into what's happening during the day. I'm not a gamer but I have wasted many an hour playing Doom or solitaire. Once I started playing seriously with Photoshop I found I had barely scratched the surface. There are many websites that have wonderful lessons on Photoshop. I guess the one thing I haven't talked about yet but obviously the computer is wonderful for emailing people, communicating with other people. (Fred)

As you can see, Enthusiasts see many different uses for a single ICT across many different areas of their lives. Fred uses his computer for work (teaching and presenting), leisure (entertainment and gaming), community use, and communication with family and friends. When you ask how Enthusiasts use a single device, they typically name every major life context and how they use it in those contexts, often speaking for upwards of 10 or 15 minutes about a single device. This is in contrast to other types, such as Practicalists (Chapter 3), who tend to see a single device as being primarily for one life context (such as primarily for work rather than family).

Another unique trait of Enthusiasts is that they are often the "technological change agents" in their work. Their views on their workplaces (even those who work in IT professional roles, such as Harry and Fred) are that their employers are simply not using enough technology or not using the technologies they have to the fullest extent. Although Alice did not work directly in IT (she worked as a home healthcare nurse), she was often pushing her office to adopt more technology:

> I think IT is great. I would encourage my home health nurse office to do more of it. I would prefer to email a lot of stuff. I'm not a "going to the office person." I'd rather email paperwork. But because of HIPAA they're antsy about privacy issues, so I can't even email something that has a client's name. I find that absolutely ridiculous seeing as how other places such as other doctor's offices go electronically between each other etc., etc. (Alice)

As Alice states, she wishes her office would adopt more technologically savvy ways of doing things, such as Electronic Health Records (EHRs). Enthusiasts tend to not be exposed to or adopt a technology because of their work life. Instead, they are the people who are bringing new technologies and ways of doing things into their work environments that they first encountered elsewhere.

For instance, Alice (as mentioned previously) had adopted a smartphone in her personal life and then started using it for work. She shared with many of her coworkers the advantages they would have if they also adopted such ICTs.

Just as in the introduction of ICTs to Enthusiasts' lives, relationships are very important in the use of ICTs. Enthusiasts, as mentioned before, tend to bond with others over the use of technology. Harry found that he had developed a very close relationship with one of his daughters because of her technical work. Harry's daughter Katrina reflected that technical discussions had brought them closer:

> Information technology gives us (my father Harry and I) something else to talk about and some common ground, so he talks about what's new, like the newest cell phone. I've shown him my computer or my work and we'll talk about it. Whereas my mom might not know what I'm talking about or might not care just because she doesn't know about it. For my job I edit video on a computer a lot. Harry has started to do that, and he'll have suggestions for me, and so we talk about that a lot. I call him for questions more than not. He's sort of known as the computer guy for the family. My brother, my sisters, my aunts and uncles, they all call my dad for computer problems or computer suggestions. He is the technical support guy [in our family]. [But] they're not usually the oldest. (Katrina about her father Harry)

Everyone around Enthusiasts easily recognizes their love of technology. As Katrina spoke about, because of Enthusiasts' timely and extensive knowledge of ICTs, they tend to become the technical "help" people for less technical family members, friends, and coworkers. It is interesting to note that Katrina mentions how "every family" has such an unofficial technical support person, but it is not often an elder. This underscores that Enthusiasts are indeed experts in technology – not simply experts in technology for their age. In fact, because of their love of technology, many of their friends, family, and coworkers expect them to be knowledgeable about all forms of technology and devices:

> Patty (coworker, friend) thinks I should know all the technical answers if she has a technical problem and if there's some issue with like a website or something she'll call and ask me. I mean her expectations are that I know about computer hardware and computer technology, which is pretty much everyone's expectation. (Harry)

As a result of their love of and for technology, Enthusiasts tend to, on the whole, embrace this role as an unofficial technical support person. Occasionally, they may be irritated by this role, but it is important to note that their expertise transcends their age, and they break many of the stereotypes the average person has about older adults' ICT use.

Enthusiasts want to expand and stretch every ICT they use over every context of their lives, so their surroundings reflect their desires to constantly be in touch with their devices. Their homes and workplaces have ICTs prominently placed to facilitate their use in every life context (family, work, leisure, and community).

ICT Display

When one walks into the home of an Enthusiast, one is confronted almost immediately by some form of an ICT. Enthusiasts love using technology in every aspect of their lives; to be used often, ICTs must be accessible and readily available.

It is important to note that Enthusiasts make no attempt to hide their ICTs, unlike some of the other user types. They tend not to have cabinets that close over the television or over their computers, hiding them and making them less accessible. Making any technology inaccessible would prevent use, and Enthusiasts enjoy using their ICTs – constantly. They are quite proud of the ICTs they own, and such ownership is an important part of their identity. Technologies form the centers of their rooms, with furniture organized around their televisions and stereos. These ICTs represent the "digital hearth" of their homes – instead of their furniture being orientated to the fireplace, it is orientated toward ICTs (Flynn, 2003). When walking into their home, the first thing that a person often notices are objects such as computers, televisions, stereos, and phones.

As can be seen in Alice's home (Figure 1), one is immediately confronted by several pieces of technology – her television, stereo, and video players are

Figure 1. Alice's Living Room, Seen From her Front Door.

readily available and feature prominently in the room. They cannot be missed when you enter her front door, despite her ability to arrange them in multiple locations in this room that would not be in someone's direct line of sight upon entering her home.

You will likely also notice that the cabinet is open, despite the fact that it shuts, hiding the DVD player and VCR. When asked if she normally shut this cabinet, she said she shut it only rarely. She commonly leaves it open, as once she shuts it, she tends to open it up almost immediately again. Alice commented that she "liked the appearance of technology" and while she was happy that designers were thinking more about how ICTs looked, she was often frustrated by the "ugliness of the cords." While she saw no need to hide the technologies themselves, she was not happy with the unsightly cords that came with them.

The armchair, barely visible to the right of Figure 1, is where Alice leaves her laptop when it is not in use. Typically, when she is sitting on the chair she is using the laptop with the television or stereo on. Enthusiasts enjoy using multiple ICTs at once, as Alice shared when I called her one afternoon to arrange her next interview:

> I was sitting here and working on a slideshow. Give me a minute to turn the television down [...] I'm trying to manipulate several photos in Photoshop and have a bunch of programs open on my computer: I often work on several things at once with the TV running. (Alice)

Such multi-ICT use is very common with Enthusiasts. Hence, their environment tends to cluster many ICTs in the same location. Since Enthusiasts love using ICTs they fill their house with technologies to allow frequent use. Unlike some other types, which may have a single computer in their office, Enthusiasts are likely to own multiple computers spread throughout their home, allowing use whenever and wherever the mood strikes (which for Enthusiasts is often):

> So, my computer upstairs is a desktop. The laptop here [in the living room] I use for email and some things like slideshows and stuff like that. I have a little one that I've got in the kitchen and I use that when I travel and basically, it's my cooking computer. I've got all my recipes on it, which is great. I call that my kitchen computer. (Alice)

These computers all serve slightly different purposes. But as a typical Enthusiast, Alice having multiple devices does not result in "extras" being stored away or unused – but indeed every device is used, almost daily. Having computers in three places in her home allowed Alice to "play" with a computer wherever she may be, and whatever she may be doing, whether television watching or cooking.

Fred, who lived independently from Alice (despite their romantic partnership), keeps his technology in his "command center," an attic space in which he spends nearly 90% of his waking time at home. When one walks up the stairs into Fred's command center, one immediately sees his computer, several printers, and various other technologies.

Much like Alice, Fred's technology takes center stage. In addition to a computer, printer and stereo, he also has a large amount of digital photography equipment, including cameras and lights, as well as a high-resolution scanner and other equipment. He would like to air condition this space, as well as add a bathroom, as this would make the space more inviting. He states that no matter the conditions, however, he would choose to be in this space with his technology:

> If I had a bathroom I could spend all day up there. I've been thinking about putting one in. There's a sink and a refrigerator so I keep soda in there, but it's generally comfortable up there. Now come mid-July and August even with the air conditioner, sometimes it isn't that great but it's livable. I just sit in my undershirt when it gets too hot. The air conditioner does the job and in front of it is a fan to blow the cool air around. So, yeah, it's livable up there and during the winter I actually have this heater thing, that's been an amazing. I just got it last winter. I'd use the space either way – too hot or too cold, but the reality is I'd be up there whether I had heat or cooling. But heating and cooling is nice to have. (Fred)

For Fred, the use of technology takes precedence over physical comfort.

Enthusiasts' homes contain ICTs in almost every living area possible, including the bathrooms. The mobility of technology makes it easier for Enthusiasts to use ICTs in such locations. Alice commented that, "Before I got rid of my landlines because of the cost and just went to the cell phone, I had phones everywhere in the house. I even had one in the bathroom!" For Enthusiasts, even the call of nature cannot separate them from their "fun toys."

Enthusiasts love their technologies, and their homes reflect this love and the central place ICTs have in their lives. Their placement of ICTs throughout their homes as the focal points of rooms demonstrates the centrality of these devices in their daily lives. This display is quite different from other user types, some of which place ICTs in specific areas (Practicalists) or may attempt to hide them (Guardians).

ICT Meaning

Technology tends to be a common thread that runs throughout Enthusiasts' lives, as Fred speaks about:

> Two things have really shaped my life: photography and technology. I've been a photographer since I was in elementary school. I like photography, even dickered with the idea that maybe I would in college do something artistic. I was probably in 5th or 6th grade my father bought me a camera and I've been fascinated with cameras and really got into photography. Now even though I'm not an engineer, my dad would occasionally bring home these really fancy radios and whatever and so I got interested in electronics and then in college I got even more into it. If you wanted high fidelity it meant buying a kit and a soldering iron and you put it together. So, I've always been interested in that kind of stuff. Technology got me jobs, and photography was always there. There are two themes in my life: technology and photography. (Fred)

It is important to recognize the multiple meanings that technology has for Fred, mirrored in all Enthusiasts. In childhood, it was an interesting hobby (leisure) and in adulthood it became a profession (work). However, ICTs go beyond just simply being work and/or leisure to Fred. Technology is a lifelong *passion* and, much like photography, was something that was always there for Fred, no matter the circumstances. Returning now to the quote we read at the beginning of the chapter, we can see how technology is more than just devices to complete tasks or to maintain relationships, but instead represents a much deeper set of meanings to Enthusiasts:

> Oh, I love technology. I have ever since I started using it way, way back when. But I just fell in love. I love everything from the word processor to the projector to making film strips…I was just in love. I've enjoyed the advantages of this kind of thing ever since they started making it available. I'm like a little kid in a candy store. I love to play around with everything I just love this stuff. Love it. Love it! (Fred)

These feelings were echoed by all the Enthusiasts, who when asked to describe their feelings toward technology responded with words such as "love," "toys," "fun," and "play." Harry describes his own attitude toward technology when reflecting back upon his relationship with photography:

> I've played with digital cameras. I've had a digital camera since the beginning – to use here at work – because we had digital cameras going back to the very first one […] And they are great fun. (Harry)

It is important to note that Enthusiasts have a great sense of attachment to their ICTs. They often feel nostalgic toward their devices and software. Fred speaks of his love and sense of nostalgia toward one program he enjoyed using, called Sidekick:

> I had started using years and years ago, an application called Sidekick. And Sidekick was a three-part program: it had a database, it kept records, mostly names and addresses but it was actually a little database, you could keep just about anything you wanted on it. It could be a record of names and addresses and you could have separate databases. These could be my personal friends, these could be galleries, and they could be just vendors that I use. It was very easy to look up information. It had a calendar with it so you could keep records of appointments. Up until Windows Vista it worked fine. Now I'm piecing together over four pieces of software to get anything near what Sidekick did and I still don't have the functionality. I really miss Sidekick. I considered going back to Windows XP, but that doesn't really make sense. I wish they come up with a new version of Sidekick, I really miss it. (Fred)

In Fred's language, you can easily see that he was attached to Sidekick not just because of its functionality (although that was a major part), but because he deeply enjoyed using it. Enthusiasts tend to form an emotional attachment to their technologies and, therefore, regret losing more than just their function: they miss the technology and the experience of using it. While the other user types discussed in this book may speak toward the functionality lost, the emphasis would not be on "missing" the technology – an emotional word that Fred chose purposely, because the loss of Sidekick was greater than just a loss of function – but a loss of enjoyment.

Despite their feelings of nostalgia, Enthusiasts are the most likely type to update their software and devices because they want to try new things and obtain the latest functionality. However, like all users (regardless of age) they struggle with updates that result in more dramatic changes to the user interface or functions. As Alice shares, she had difficulty using a newer version of Photoshop and kept using her older version on her older laptop until it failed:

> I find the longer I have a piece of equipment, especially the smartphone, the more I find out about it [...]. So, I'm still fighting with Photoshop, but I'm getting better at it. For about a year until the old laptop died I was sitting down here using the old version of Photoshop, which was like three or four down from the new one, and putting my finished product on a thumb drive and going upstairs printing it because I could do in 10 minutes what it was taking me two hours to figure out on the new [version]. The fact that they change these things and end up confusing people is difficult. Don't get me wrong – I'm not going to

> stop using Photoshop, but I'm hesitant to upgrade once I know how to use a piece of software. (Alice)

As you can see, Alice is determined to use the newer version of Photoshop, but is struggling with it. Unlike some other types, who may be more likely to quit using software or hardware they find challenging (or ask someone else to complete the task) Enthusiasts tend to keep trying. They are the most likely to update of all the types, but they weigh any potential update by considering the increased functionality against the time it will take to learn the new version.

Enthusiasts realize that their love of "things with cords" (Alice) and technology in general is quite different from most of the population. When asked to talk about their technology use in comparison to others their age, they often said that they were much more advanced. Alice shared the following:

> I'm quite high tech for my age group. I do know I am quite high tech but there's a lot of people that are more high tech than me of different ages because that's the circles that I run in. I prefer hanging out with people who are tech savvy, but I know a lot of people my age would think that is strange. In fact, a lot of people younger than me would think my level of tech savvy is strange! (Alice)

Alice's comment recognizes that her heavy use and, in particular, her enthusiasm for ICTs are quite different from those of the general population. In fact, all the Enthusiasts in the study labeled their use as different not only from other people of the same age, but also from even people who were younger. Fred recognized that many people likely would think that his love of ICTs was "weird," unless of course, they also loved technology.

Enthusiasts: The Technological Evangelists

One can think of Enthusiasts as evangelists for ICTs. They encourage people to try new ICTs, surround themselves with others who use and love technology, and push their workplaces to incorporate more ICTs into their processes and procedures. Some key takeaways about Enthusiasts include:

- Enthusiasts have a lifelong love of technology that began in childhood and was encouraged by mentors.
- They are the most willing of any of the user types to try a new technology.
- Enthusiasts learn about new technologies through their own research (technical blogs and articles) as well as relationships with other technically savvy friends and family members.
- They place ICTs in prominent places in their homes and have a strong preference for technologies that are beautiful.

- Technologies are fun toys to the Enthusiasts, so to appeal to this user type one should emphasize the fun nature of an ICT.

Chapter 3 explores the Practicalist ICT user type. While Enthusiasts love ICTs as fun toys and are constantly looking for new ways to use them, Practicalists tend to view ICTs as tools that fulfill a certain purpose in one area of their lives.

Chapter 3

Practicalists: The Technological Tool Users

> I see my cell phone like a hammer. I take my hammer out of the toolbox when I need it. I don't carry my hammer around with me every day all the time. I use the cell when I need to make a call, just like I use my hammer to drive a nail. I don't walk around carrying my hammer for someone else to use it, it's mine [...]. These technologies they are just tools. (Boris)

Tools. Function. Purpose. Task.

When Practicalists speak about using an ICT, they focus on the technology's usefulness, function, and purpose. Practicalists do not stay up all night playing with a new gadget, exploring all the nifty new features. They will not be the first in line at the store to buy the newest technology or be well-versed in all the latest updates. Practicalists view ICTs as purposeful tools meant to get a job done or complete a task. They see individual technologies as fulfilling a specific purpose/function in their lives.

Practicalists often are (or were) involved in paid work that involves heavy ICT use. Work is an important point of introduction for Practicalists, as it is through work and its associated tasks that they most often encounter new functions and technologies. They place ICTs in their homes in specific functionally dedicated areas: computers belong in offices, televisions in dens, and landline phones are placed wherever they will get the most use. ICTs are seen as tools that serve a specific and distinct function in their lives, be they tools for family, leisure, work, or community.

Formative Experiences

Practicalists, as a group, do not share the fond childhood memories of ICT use that Enthusiasts speak about. Instead, Practicalists' foundational technological stories often begin with their work careers. As a group, Practicalists have a large diversity of career trajectories, with individuals involved in blue, white, or pink collared positions. Some older adult Practicalists are retired and some work full or part-time. They are also an educationally diverse group, ranging from high school graduates to doctorate holders. This user type credits their early work lives for exposing them to technology, such as when Jack speaks about his career trajectory:

> When I got out of high school, I went to work for a radio and TV repair place. I was in electronics repair. I went to a technical high school and then I went into the Army. When I got out of the Army Reserves, I went to work full time for Western Union. I was an electronics repairman. Then I went to work at the Air Force Base where we had a computer line for the Department of Defense. I worked there for 25 years and then I left there and retired and got a pension from them. I worked at The Mall as an electrician when they were building that in 1989. Then when that got built I was all set for full retirement, but my wife said they needed a plumber, an electrician, and a carpenter up at the Catholic Diocese, so I went to work there. I retired from there too. I can do just about anything, plumbing, electrical, carpentry [...] I'm mechanically inclined and electronically inclined. (Jack)

Despite the centrality of technology to his career, Jack lacks the passion for ICTs that Harry, Alice, or Fred (all Enthusiasts) exhibited. Though Jack's early working life (including technical experience and military service time) was similar to Enthusiast Harry's, his philosophy on ICTs is dramatically different. What resulted from Jack's early adulthood exposure was not a passion for ICTs as fun playthings, but rather a "practical" and "functional" perspective. Practicalists do not talk about a "love" for ICTs or share stories of how ICTs dramatically shaped (or saved) their lives as Enthusiasts do. They do not get excited about all the ways they use (or potentially could use) an ICT. Instead, they see ICTs in a very pragmatic, purposeful, and functional way: as tools.

Practicalists, unlike Enthusiasts, do not point to important technology mentors (family or friends that encouraged use) nor do they have fond memories of "tinkering" with technology (being encouraged to break down, rebuild, and play with ICTs). When asked about their childhood experiences and how these have shaped their relationship with technology, Practicalists often dismiss their childhood technology encounters as not being influential or memorable compared to their working careers.

Introduction to ICTs

It is through work tasks and processes that Practicalists are often introduced to new forms of ICTs. Practicalists' work often has a strong technology component. Belinda's former work as a library science educator and currently as a professor meant that she worked extensively with ICTs. Cleveland, who prior to his retirement worked as an executive for a paper sales company, traveled frequently and used many earlier mobile technologies. Boris computerized his construction business billing and records in the early 1990s.

In their work, Practicalists take a functional approach to learning ICTs. Their learning is purposeful, and they see learning not only as a work task, but as work itself (as opposed to being play or fun). Belinda speaks about how she is constantly trying to learn to use new software and applications in her teaching:

> I constantly am trying to push myself to use new tools. What I am really trying to do is to see how I can use these in the context of librarianship. I try to figure out how these new tools can be used for research assignments for kids. I don't go out and search for these technologies, but if someone lets me know they are out there I try to figure out how to use them. (Belinda)

Practicalists do not approach their ICT use with the pure, almost childlike, wonder and excitement of Enthusiasts. Instead, Practicalists approach ICT use as a focused and driven pursuit to find out how to use the technologies available for the task at hand; they do not play with them. Belinda's quote captures this driven and focused examination: she does not seek these technologies out, but fully explores their potential uses for the educational context when she is made aware of them.

While Enthusiasts tend to push their workplaces to use existing ICTs in new ways and to adopt new technologies, Practicalists are more likely to be encouraged by their workplace (be it their peers or supervisors) to begin using these same ICTs. Cleveland speaks about how his experience with cell phones evolved through his work exposure:

> I first became aware of cell phones around 1990. The owner of the company got wind that this was the coming thing so he got a cell phone. The only way I can describe it is it looked like a WWII combat walkie-talkie. This thing was about 12 inches long and it weighed about 6 pounds. It had a keypad on it and you held it up and it was limited coverage. We used that for maybe a couple years and maybe he got another one. Maybe there was one or two that were being used, interchanged amongst the group in the office. But they were just terrible, erratic reception, etc. Then I went to Asia and I can remember spending a few days in Taiwan. It was a very crowded city, very, very busy; very, very hectic; very, very modern; very, very high tech. Everybody was walking along the street – I want to say almost everybody walking along the street was talking on a mobile phone. We were just flabbergasted that this technology that we thought was so difficult to get adjusted to was so common place in Asia. We felt like we were living out in the boondocks. So, we came back and cell phones became a little bit more common place, a little bit more

> available, and a little bit better reception [...] So, the cell phone activity in Europe and Asia was very common place, but the phones that we were getting here in this country weren't usable there at all. You'd have to go over there and if you wanted to make a cell phone call you'd have to buy a phone, which I did a few times and that worked out fine. I realized after a while that we're way far behind in this country in terms of global phone technology. (Cleveland)

For Cleveland, his experience with the cell phone mirrored his use of many technologies: his first computer and later laptop were provided through his office. In his discussion of the cell phone, Cleveland speaks toward the practical use of this technology: its reception, functionality, and usefulness: important aspects of an ICT to any Practicalist. Comparing Cleveland's (a Practicalist) discussion of the cell phone with Fred's (an Enthusiast) discussion of the television (first introduced in the previous chapter), you will note that both discuss the limited functionality of the first models. Whereas Cleveland mostly focuses on usability, Fred mostly focuses on the fun:

> I remember the first TV I saw [...] it was a little TV screen, in a big box. It was black and white. In the '50s there were only three channels in my city. After 11:30 at night the only thing on was a test pattern. That was the early '50s. It was amazing and so nobody on our street had a TV and then the one kid I hung around with on the street their family got a TV. We'd go down there and watch TV and they had the fights on Friday and that was about it. But it was amazing. Later they had movies on TV, you could watch the news [...] it was great! (Fred)

The meanings Cleveland and Fred ascribe to these ICTs are vastly different and reflect their different user types. Cleveland focuses almost solely on the functionality in his quote about cell phones, while Fred focuses on the wonder and excitement of television. Fred, as an Enthusiast, is amazed and excited by the "fun" of a new technology. His excitement nearly jumps off the page. Cleveland, on the other hand, is impressed by the functionality of the technology: its usability and ubiquity. While Practicalists are not eager to adopt new ICTs because they are "fun toys," they are eager to adopt new technologies that they believe will benefit them functionally: they are personally or professionally useful to them. For instance, Boris, who ran his own construction company, first adopted a computer to help with typing letters, billing, and general recordkeeping for his business:

> The first computer I had was for the business – it was an inexpensive tool. You could do certain things with it, you could type

> a letter and you could print from it. But it was at least that 20 or 30 years ago. (Boris)

As the sole proprietor of a small construction company, Boris was a relatively early adopter of computers in the 1990s for his home business. To him, the computer was a tool that would help him with inventory and other business work. Practicalists, like Boris, are neither overly impressed with nor intimidated by technology – they simply want to use the things that help them to "get stuff done" (Jack). This pragmatic view of ICTs as tools can impact Practicalists' exposure levels. Because ICTs are so closely tied to work in Practicalists' minds, their overall exposure is closely aligned to their workplace's ICT status. Practicalists who are in workplaces that possess advanced ICTs, and are in positions where such use is required or expected, tend to be well versed in advanced ICTs. Those in workplace environments that lack ICTs or that are/were in positions where using such ICTs was discouraged tend to have much less exposure. This can be observed for those Practicalists who held high positions in their organizations as executives and retired before executive-level employees were expected to complete their own computer tasks. For instance, Dan, who held a high-level executive position in a global non-profit agency, relayed that early computers were seen as being advanced typewriters for typists. At one point, he even asked his mentor if he should be learning to use computers and was told such learning was beneath his position:

> I remember I was running a major project in Africa and I didn't have computer skills. This was 40 years ago or so – in those days you had lots of other people who did computers for you: you had typists etc., etc. But even when we started with computers typists just switched over to computing. Now that's changed, now you're expected to have your own skills. I remember talking to a man who was a senior administrator and had retired from some agency in California. I said, "Allen do you think I should be learning the computer?" […] He said "well, you know Dan at your level, no; you shouldn't be using a computer. Other people should be doing this for you." (Dan)

Throughout his career as an executive, Dan believed that computers were simply tools, albeit not tools that he needed to learn for his work: they were the tools of typists and administrative assistants. As computers became more widespread in offices, Dan sometimes felt that he should be using computers, but it always seemed that someone else in the organization would preemptively prepare documents and presentations:

> I always had people build [the presentations] for me. Maybe that's a disadvantage if you have other people do it. If you're at a certain level in an organization you have other people do things.

> Then you haven't really learned [...] I could give a great PowerPoint presentation but somebody else had put it together. I'd really like to learn PowerPoint. (Dan)

Even though Dan previously held a high-level executive position, he found himself struggling in retirement to learn to use office software due to this lack of exposure. Dan wanted to start his own consulting company and was also serving as an advisor to many doctoral students who were studying international development. He felt he lacked many of the computer skills needed to be successful and was in the process of learning to use many office applications, both through lessons from his wife and by reading books.

It is often assumed that an individual's education level and socioeconomic class correlates strongly with their ICT use, proficiency, and exposure in elderhood (Czaja et al., 2006; Peral-Peral, Arenas-Gaitán, & Villarejo-Ramos, 2015; Pick, Sarkar, & Johnson, 2015; Zhang, Grenhart, McLaughlin, & Allaire, 2017). However, despite Dan's high level of education (a PhD holder) and high socioeconomic class (lower upper class), his exposure levels were quite low. This turns our assumptions about the impact of education and socioeconomic class on its head. It is not simply exposure to ICTs that leads to better skills, but the direct use of them. Indirect exposure – where ICTs are available in an office but people do not use the technology themselves – does not result in greater skills. While work is a key point of introduction for Practicalists, it is important to note that not all Practicalists are personally introduced and taught how to use all ICTs during their careers.

Practicalists often purchase their own versions of ICTs for personal use. They tend to readily understand the features and characteristics they want and need in a technology, and have high awareness of the ICTs that are available. Boris relays how he tended to purchase technologies:

> If it's something that is going to make your life easier, more pleasant, and you can afford it then you buy it. For years we had normal regular [CRT] TV sets. They worked, they weren't great, but they worked. Two or three years ago both of us agreed that what we would do for Christmas instead of buying a bunch of dumb stuff we would buy a [LCD] TV set. We researched them a little and talked with a bunch of people that had them. We went and looked at them in the store and bought one. Well, then I had a chance to buy a small flat screen for the bedroom at a pretty good price and again I checked with my computer guy to make sure it was a good deal. (Boris)

Like Boris, Practicalists often do an intensive amount of research on any potential ICT purchase. For Practicalists who are in close relationships with Enthusiasts, they often find these relationships' important points of introduction to new ICT forms and devices in addition to their work lives. Dan's Enthusiast wife often purchased new forms of ICTs for family use:

> Probably the only reason I have anything is because my wife has all of this [ICT] stuff. For all of this stuff I don't know what I would have if it weren't for her. Maybe I would have it all and maybe I wouldn't. It's her influence. She's a genius in this type of stuff; I just try to figure out how to use it. (Dan)

While Enthusiasts actively seek out new forms of ICTs in their everyday lives, Practicalists tend to become aware of new forms most often through their work relationships. Since work has an important role in the introduction to ICTs, Practicalists' have very diverse skill levels when it comes to ICT use, depending on their type of work.

ICT Use

Practicalists ICT use patterns tend to reflect their belief that ICTs are function-specific tools. They often use an ICT in only a single life context, such as their family, work, *or* leisure lives. This is in contrast to Enthusiasts, who seek out all the possible applications of a single ICT across all areas of their lives; for their family, work *and* leisure lives. Practicalists, in contrast, are much more specific in their perceptions of an ICT being intended for one area of their life or another. They tend to see television as a leisure device for family, cell phones for work, and computers for work or activities related to maintaining their household:

> Primarily I would say the computer is for business, business records and research and things. There is some personal use; I've got a couple games on there I play. I look at the news and the weather, but mostly its business that's in there. (Boris)

For Practicalists, their ICT use is driven by the applications they see for the device or software in their lives, be it professional or personal. When Boris speaks about his computer, there is "some" personal use, compared to being "mostly business." This function-specific description of his computer is much different from Enthusiasts'. Even though Boris does use his computer for leisure (gaming, the weather forecast, and news), he views this use in a functional and purposeful way to fulfill his leisure needs. Even when playing games, the computer is a leisure tool, not a toy, to Practicalists.

Like Boris, Belinda speaks about the precise functions she uses an ICT for, in this case, social media:

> I'm a learner. When you're talking about social tools and web-based presentation things and a lot of the new applications I am

> fledgling. I push myself to use them. There are two ways I think about using them in my work. One is for my own personal work communication uses. I'm still finding a comfort zone with that because I'm not a real public person in a lot of things, and I prefer to write an article than to do a blog. So, I prefer to have my thinking done rather than just lay it out there in process thinking. The second way that I think about the use of particularly the new social media tools and the new applications is in terms of student research. This second way is to figure out research products that allow students to communicate their research and what they have learned in innovative ways. I think about tweets of George Washington before the battle of Valley Forge, or of all the soldiers on the night before the battle and how could kids capture the research they've done about what it would be like to be a soldier at that time. As a librarian, you collect those tweets to give an overall picture of what it was like to be in that situation the night before a huge battle. I always think about how it can be used by students and if I think it would be useful for me professionally. (Belinda)

Belinda's thoughts on using social media focus on its application in a single life context: her work. She sees two prominent uses: to enable her to communicate (and this communication is focused professionally) and how it can be used in teaching. Like Enthusiasts, Practicalists constantly see themselves as learning, but unlike Enthusiasts, what Practicalists want to learn is targeted toward a small range of function. They do not seek ways to stretch the use of a single ICT into every area of their lives. They are happy, instead, to have every ICT have a single set purpose:

> I don't regard the Internet as a toy; I use it for work, for very strategic finding of information, for getting something done. So, I don't just search, I don't just roam around on the Internet [...] I don't have time for that, and I don't want to do that. (Belinda)

Belinda discusses her use of the Internet as "strategic" and purposeful. She directly contrasts her use to someone who is "playing" (or using the Internet for fun). It is important to note that Belinda was very much a power user of the Internet: She taught online courses, maintained social media accounts, and used a plethora of common ICTs. In many ways, her skill levels rivaled several of the Enthusiasts. However, Belinda's perspective and use were quite different from Enthusiasts: she focused on tasks and function, not fun and play. ICTs are not toys to Practicalists: they are tools. While Practicalists can be highly skilled ICT users, their skill set may not be as broad because their use is targeted to a specific purpose. Practicalists know how to use the tools they use frequently, but unlike

Enthusiasts, they are not apt to explore different ways to use the technology in different contexts.

This sense of purposeful use, and this view of ICTs as tools, impacts the Practicalists' use in other ways. In particular, Practicalists believe *they* should be in control of the ICTs they themselves own. As their owners, *they* control how and when they are used. This use is for their own convenience – and not for others'. Most older adult Practicalists, for instance, maintained a separate landline, which was their primary form of short and long-distance voice communication. Cell phones were seen as personal tools to only be used at Practicalists' discretion – and particularly – when Practicalists' had a desire or need to use it (most often to make an outgoing call):

> I'm not big on cell phones. It's the matter of necessity. I have one only in case I get stranded somewhere or had a problem. Half the time I don't even take it with me. I guess I keep it turned off because I've got a couple friends and their phone rang more times in the day than you could count. Half the day is shot because of the cell phone. It's like what did you do before you had a cell phone, somebody waited to talk to you. I don't want to be bothered on construction jobs. Half the time I couldn't hear the phone anyway because I'm using a piece of equipment, saw or sander, or router or something, I'd never hear the cell phone. I don't turn mine on unless I want to make a call. (Boris)

These feelings were echoed by all the older adult Practicalists in the study. The cell phone, as Boris states, is his. He keeps one primarily in case he is stranded in his car or runs into difficulty. He normally keeps it turned off unless he needs to make a call, as that is when it is most useful to him. Dan expressed a similar sentiment:

> I normally keep my cell phone in the car. I think it frustrates other people sometimes because I don't hear it if it's in the car. In this house I always use the home phone, and so it doesn't bother me not to have the cell phone in here but it probably bothers a few other people. My wife says, "Why don't you take the cell phone with you?" I forget to take the cell phone out of the car when I go someplace else. (Dan)

Dan hints at some of the conflicts that can arise from Practicalists' view of ICTs as tools for their own use, particularly when they are communicating with others who do not share their user type. Many of the secondary participants echoed this frustration. Peggy, who was one of Belinda's friends and a former coworker, explained that she often would tease Belinda about not answering her cell phone:

> Belinda does not text at all. Do not text her. If you need to call her on her cell phone, you have to ask her to turn it on; she has to plan it to receive your phone calls [...] I keep teasing her, I said "why don't you text and I'll text you and you can receive it and then you'll know I'm calling you." She just gives me one of her looks. *[laughing]* (Peggy on her friend Belinda)

Boris and Belinda keep their cell phones off because, in their minds, this facilitates their own use of their ICT tools. Since Boris views his cell phone as a tool for emergency use; he keeps it off, as he is not constantly in need of an emergency tool. Since Belinda feels that the cell phone is her tool for work; she similarly keeps it off, unless she has a work-related call to make.

For Practicalists, all of their ICTs are life context and purpose specific. Although some ICTs may cross life boundaries (for instance, Jack paid his household bills and donated to community organizations online), the number of contexts each ICT is used for is seen as very limited. For Practicalists, it is easy to classify any device into primarily being used for either family, work, community, or leisure. Their ICT use does not often cross the boundaries of these life contexts.

ICT Display

Practicalists' apply their view that ICTs are tools used for a specific context in how they organize their homes. Computers, which are primarily used for work, are placed in offices. Televisions, which are seen as instruments of leisure, are placed into living rooms or dens. Phones are placed strategically throughout the home where they facilitate different kinds of conversation (work, family, or leisure). Placing a television, a device for leisure, in their kitchen or office is akin to leaving gardening or woodworking tools in the same spaces. To a Practicalist, neither makes sense. Belinda speaks about her placement of her computer, television, and printer in her den/home office:

> I put my computer here [in my office/den] because I love to look out; I love to have a backyard to look at. I spend all my life on my computer, so I need it to be comfortable and facing out. It's convenient for me; this is sort of a base. I put the television in the room because that's kind of the den, that's where I would go to relax, and the living room is more for entertaining people and conversation. Why would you have a television there? (Belinda)

Practicalists often own many of the same ICT devices as Enthusiasts; however, their view that ICTs are tools that are obtained and used for specific purposes leads them to organize them in their homes in very different ways. Whereas Enthusiasts place ICTs in areas where they will get maximal and nearly constant use, Practicalists place ICTs in areas where they will fulfill a specific

Figure 2. (a) Boris' Computer Room and (b) Boris' Entertainment Room.

function. Boris separated his open floor plan log cabin into both a separate office space and a separate entertainment space (Figure 2a and b).

Often, Practicalists tend to label rooms in their homes by the technologies they have within them. For instance, Practicalists will often have "computer rooms" or offices, "television rooms" or dens, and occasionally ICT-free spaces they use for other tasks, such as sitting rooms or reading rooms. Belinda's quote emphasizes that ICTs have their designated space and this space is based upon their function (be it work, leisure, family, or community use). The room becomes named after the primary activity that occurs in that room – be it computer use, television use, or communicating with friends, etc. This reflects the Practicalists' view that ICTs are tools, and tools are used in a designated workspace:

> We used to have the computer in the living room. The problem with that arrangement was we couldn't get a sofa and chair big enough out here. If you had more than one or two people you couldn't have a conversation because of the lack of seating. We decided that we're better off moving everything and bringing the computer out here. This is my office space. This is my business space. I just got everything to do with business pretty much right here so that I don't have to go anywhere. I keep my personal files in here, but I also keep other stuff to do with the business here like printer paper, ink cartridges, whatever. This is where I do my business. We put the TV in the other room and that is our entertainment room. (Boris)

As Boris speaks, one can see how he readily organizes his space around technological function: the television is in an area used primarily for entertainment, the computer in an area used primarily for work, and there is also a separate seating area for having a conversation.

Jack relates how he and his wife use the space in their home:

> We put the computer in here because this is a good little work area to put a computer in, like an office. We have a TV for the grandkids to use in the front room, and this is the TV that my wife and I use when we are downstairs in the living room. We didn't put the TV in the office because that is where the computer belongs. When we use the computer, we're in there. When we use the TV, we're out here. And if the [grand]kids are watching, they're watching a kid's program, so we don't watch that. They have their own TV for that. (Jack)

As Jack states, when they are seeking to use an ICT, such as the computer or television, they enter that space, use that ICT, and when completed with that task, they exit that space. This was observed over and over again in the study. If a Practicalist owned a laptop and their spouse or another member of their family moved it from the "computer room," the Practicalist would pick up the laptop and return it to the "computer room." Cleveland, for instance, commented on how Mary (his wife, a Socializer) would often move their laptop to the upstairs living room, while he preferred to work on the laptop in the basement office. During my interviews with Mary, he would often retrieve the laptop from her chair-side table and return it to the office. Many Practicalists reasoned that their use of ICTs should be self-disciplined, and locating these ICTs in task-specific rooms helped them to move into other activities, when completed with the task at hand:

> I keep the computer out in the office because it allows me to only use it when I want to – otherwise I can read or do something else. (Dan)

Unlike Enthusiasts, Practicalists do not see their ICT use as "fun" or "play" but rather the work of using a tool to accomplish a task. Therefore, part of their reasoning for maintaining these ICT-specific spaces is to facilitate their efficient use of ICTs as such tools.

The Practicalists of the Lucky Few generations tended to view cell phones as their own personal tools (for their convenience, only). As a result, Practicalists of this generation often have multiple landline phones to facilitate communication in each life context: be it in work, leisure, etc. Despite the open floor plan of Boris' home, a separate telephone had been installed in the kitchen, the office area, and the entertainment area, with these spaces being only a few feet from one another. When someone called during our interviews, five separate phones could be heard ringing simultaneously. Boris shared that his motivation in having multiple phones was to improve function and allow for greater usability:

> Having phones everywhere in the house just makes it handy. Those phones from Radio Shack are $20–25 and they're cordless and if I'm going to go out in the shed and work or puttering around outside I take that phone with me. If [my wife] gets a phone call upstairs and it's for me, and she calls me to the phone, I don't have to run upstairs. I can just come in the door downstairs and use that phone. (Boris)

From the Practicalists' perspective, this arrangement makes perfect sense. The placement of the phones around Boris' home reflects the various tasks he may be using the phone for: personal communication, business, or leisure.

ICT Meaning

For Practicalists, ICTs are tools, work, and functional items used for practical purposes. They tend to view each of their ICTs as having a distinct purpose in their lives:

> I use the cell phone in a very practical way, it's strictly for communicating, if I don't have the home phone I use the cell phone. I don't do anything fancy on the cell phone at all, just send some voice messages and make calls. (Dan)

The cell phone is valuable as a device for Dan not because it does something "fancy" (has advanced features) but because it allows him to complete his tasks of communication. In many ways, a Practicalists' main concern with any device or service is its functionality: what it can do and how well it can do it. They tend to be less impressed by new features or the attractiveness of a device, unless these features represent a perceived functional improvement. Practicalists tend to credit the environment around them: their work, their families, but most importantly the tasks they do as influencing their everyday ICT use:

> I think it's what you do, whatever you do, whatever your livelihood is dictates the kind of equipment, the technology that you're going to require to do your job. Or it's what you're comfortable with, whether it be watching a lot of TV or playing a lot of CD's or talking on the telephone or text messaging all day. It depends on what you're going to use it for. (Boris)

Boris touches on the importance of use for Practicalists. ICTs that have purpose have value in the Practicalists' world. Boris' comment highlights how Practicalists feel: a person's life and the tasks they perform dictate the ICTs that they use. If they enjoy music they might use a CD player, if they like drama they might watch television. His view of ICTs as equipment highlights the purposefulness of ICTs in his own life. Through its usefulness and practicality, an ICT comes to have value.

While Practicalists would never categorize their use of ICTs as being "fun" or "play," they do not believe using ICTs is drudgery or devoid of enjoyment. For Practicalists, the "joy" in using an ICT comes from fulfilling a task or doing a job. Some of their joy comes from finding new features in tools they are already using:

> I think you get set in your ways in using ICTs and you tend to do the same things every day the same way until one day you try something different and learn something new that's helpful. It kind of amazed me, I found out something that even my computer guy wasn't aware of. When I take the digital camera and put the pictures in the computer I can just hit transfer and it will transfer them. Well, one time for some reason I noticed a Canon icon on my desktop, so I clicked it. It's got every picture that I've got stored in that computer no matter where they're stored. Now if you want to print them that's handy you can just click on them you pick your paper and how many pictures do you want on a page, hit print and you're on your way. (Boris)

Practicalists, like Boris, often happen upon these new features passively. They discover them by chance or have new features pointed out to them. While Enthusiasts will deeply explore any device they own to find out every way it works (and find the exploration process great fun), Practicalists tend to avoid this experimentation and instead focus on the functionality they are already familiar with. Practicalists want technologies to be shown to them and tend to strongly prefer technological training over learning through exploration. They also like having manuals and technical support individuals to walk them through issues, as they are unlikely to try to solve technical problems on their own. Practicalists do not want to play with a technology; they want a technology to work as expected. Practicalists, however, are open to chance discoveries and welcome new uses being pointed out.

Practicalists focus on the usability, and the function of ICTs is easily spotted by those around them, be their coworkers, friends, or family members. These individuals recognize the Practicalists' serious use of ICTs as functional tools, not playthings:

> Belinda's ICT use is pragmatic. If she needs to use it she will and she'll use it well. She would never get up at 5:00 in the morning and play with a piece of technology. (Peggy about Belinda)

As Peggy states, Belinda would never "play" with an ICT, but she will learn to use it well. When Practicalists view an ICT as a tool, they see the technology much like a hammer. Returning to Boris' quote which started this chapter, we can see how ICTs are viewed as tools, and like any tool, have accepted uses:

> I see my cell phone like a hammer. I take my hammer out of the toolbox when I need it. I don't carry my hammer around with me every day all the time. I use the cell when I need to make a call, just like I use my hammer to drive a nail. I don't walk around carrying my hammer for someone else to use it, it's mine [...] These technologies they are just tools. (Boris)

Practicalists' language is littered with the term "tool" to refer to technologies. Boris, Belinda, Jack, Dan, and Cleveland all referred to ICTs as "tools." For Practicalists, the ICTs in their lives which have the greatest amount of value are ones that serve a practical purpose and have a determined function: they are simply tools.

Practicalists: The Technological Tool Pragmatists

Practicalists are the pragmatic tool users of ICTs: they are more than happy to adopt any technology they believe will serve a purpose in their lives – if you show them the technology and how to use it. They tend to take a serious approach toward using ICTs, viewing their use as more work than play. This work can be joyful, but it is a serious task, not a fun one. Key takeaways about Practicalists include:

- Practicalists have a large diversity of skills as a group, with those who were heavily exposed to ICTs in work having the highest skills.
- They extensively research the functionality of new ICT purchases.
- Technologies are placed in function-specific areas in Practicalists' homes: computers belong in home offices and televisions belong in entertainment rooms.
- Practicalists view ICTs as context-specific tools for work, family, leisure, *or* community.
- To appeal to Practicalists, one should emphasize an ICTs' functional usefulness.

While Enthusiasts love ICTs as fun toys and Practicalists see ICTs as tools that fulfill a certain purpose, Socializers value ICTs for their potential to connect them to others. Chapter 4 explores how the Socializer ICT user type views technology as connectors and, in particular, how technology is a connector between generations.

Chapter 4

Socializers: The Technological Social Butterflies

> I am the queen of texting. I have to text. I'm forced to do texting because some of my grandchildren just will not answer the telephone. They have their phones on vibrate so they just will not talk on the phone. So, if I want to ever talk to them I have to text. I'm a great texter and I know all the abbreviations. I make some of them up myself and I have them ask me what they mean. "Huh you're a pretty smart old lady" they'll text me back. I make up my own text words [...] I don't know what they say or think about me doing all this texting, but I love to do it. And I have to do it. (Gwen)

Connectors. Communication. Socializing. Bridging generations.

For Socializers, Information and Communication Technologies (ICTs) are devices, applications, and services that connect them to others. Unlike Practicalists, Socializers do not view ICTs as tools, but rather as communication bridges between people. Unlike Enthusiasts, Socializers do not love all ICTs, but instead love those technologies that facilitate and encourage socialization.

Socializers are highly involved in their communities and tend to have large intergenerational families. It is through these intergenerational contacts that they are introduced to new ICTs. Their active lives mean that they prefer mobile technologies that allow communication while going about daily tasks. To Socializers, a valuable technology is one which allows them to build a connection with others.

Formative Experiences

Socializers do not have memories of early mentoring and positive technology interactions that Enthusiasts have. They do not credit having worked in positions that had high contact (direct or indirect) with ICTs as influencing their use, as Practicalists do. Instead, what seems to have shaped Socializers is their natural tendency to be extroverted and their desire to connect with others, something that these individuals shared was a life-long trait. Technology, throughout their lives, has always represented a means of *connection*.

Introduction to ICTs

While many Socializers encounter ICTs through their work, the ICTs which hold Socializers' interest in their free time are those that serve a communication purpose. Upon retirement, many Socializers simply stop using ICTs which they consider to be non-communicative in nature. Gwen started her career as nurse and later transitioned to being an administrative assistant. Upon her retirement, she concentrated primarily on using social ICTs, such as her cell phone for texting.

Socializers' introduction to new ICT forms happens primarily through their relationships, particularly relationships with younger individuals. Mary shared that many of the things she had learned to do online were influenced by one of her five children:

> I learn a lot from the kids. I see what they use and how they use it. This past Christmas they were here and they were showing me their iPads and their iPhones. And I learn a lot from them about what you can do with the stuff. (Mary)

Socializers respond differently to being shown such ICTs than many of the other types. Whereas Enthusiasts are constantly discovering new ICTs on their own and Practicalists are often introduced though work tasks or colleagues, Socializers are introduced to new ICTs through their relationships with younger people. Socializers are eager to have others show them new technologies, but are specifically interested in being shown new ways to communicate and interact. Gwen speaks about how her children show her their smartphones and tablets, and she is constantly learning new ways to use them:

> I want whatever the grandkids have [...] My daughter had an iPad the other day. She showed me her daughters' gymnastics performance on it. I asked her "show me how to use it." I was a bit afraid at first, but she said you just touch it like this. Now I want one. (Gwen)

Like Gwen, Socializers are keen to understand what younger individuals around them are using to communicate. Socializers are deeply embedded in their community: highly involved in religious organizations, in their neighborhoods, and in charity/volunteer work. This community work brings them into contact with individuals from a wide variety of backgrounds and these contacts often cross gender, racial, generational, and socioeconomic boundaries. For Socializers, these community contexts can be important introduction points of new ICTs, often given as gifts. Socializers tend to be generous with their time in these contexts and often find their generosity reciprocated, such as in the case of Gwen:

> So, I went into church the other day, and the pastor said to me, "Some people found out your digital camera was stolen, and they wanted you to have this." He handed me a package and it had a digital camera in it. I couldn't believe it! Now I need to figure out whom to thank. (Gwen)

Gwen's digital camera was stolen at a community event and she was gifted a new camera at church. Because Socializers invest so much into their communities, these communities often want to give back to their Socializers. The gifting of these technologies is often more than a thank you, however; the Socializers' presence in these communities is viewed as *critical* and it is understood that technology facilitates Socializers' involvement. Socializers who are unconnected from their communities cannot participate in them; therefore, it is critical that these communities help Socializers stay connected.

To Socializers, an ICT that connects them to others has value, particularly if it allows them to connect to their multigenerational network of friends, family, and community members. Socializers always want to know what ICTs their young friends, family, and community contacts are using, and how they are using them.

ICT Use

Socializers value constant contact. It was easy to identify a Socializer based on one feature alone during the interview phase: interruptions. Interviews with Socializers were constantly interrupted by phone calls, texts, and individuals' knocking on the Socializer's door. Even when Socializers would silence their phones, oftentimes interviews were interrupted by individuals who knocked on the door, worried about the Socializer being "out of contact" for a few minutes. When Socializers would turn on their cell phone at the end of a three-hour interview, they were often met with a dozen text messages and/or voicemails, announced by a steady stream of pings and rings. Oftentimes Socializers would check these messages quickly before even setting up the next interview.

Socializers are the most eager of all the five types to adapt to "the young's way of doing things" (Mary). Socializers have come to believe that the only way to stay in touch with their youngest contacts is to mimic the younger generation's communication habits, adopting the same technologies. When we re-visit the quote from Gwen at the beginning of the chapter, we see that she is eager to adopt the habits of her grandchildren when it comes to texting:

> I am the queen of texting. I have to text. I'm forced to do texting because some of my grandchildren just will not answer the telephone. They have their phones on vibrate so they just will not

> talk on the phone. So, if I want to ever talk to them I have to text. I'm a great texter and I know all the abbreviations. I make some of them up myself and I have them ask me what they mean. "Huh you're a pretty smart old lady" they'll text me back. I make up my own text words. My son, of course, he's a sheriff and he's very busy so I'll text him real quick just "? R u kp" (Are you a cop?) or "?location" (Where are you?). I don't know what they say or think about me doing all this texting, but I love to do it. And I have to do it. (Gwen)

Gwen states that she both "loves" but also "has to" text to stay in touch with her youngest family members. Socializers do not enjoy using any ICT itself as a "toy" (like Enthusiasts do), but rather enjoys the relationships that ICT use helps them to build, facilitate, and maintain. It is important to note that Gwen speaks with the words "have to" much more frequently than "want to." Gwen loved having close relationships with her grandchildren, and she was keen to maintain those relationships. For her, that meant that she needed to learn to text, including learning all of the "text speak" her children and grandchildren used. She did not learn text speak because it was fun in itself (although one can see how she has fun using it) but because she understood its value in communicating with younger people. Older adult Socializers often feel that their use of ICTs is simply mimicking the use of younger individuals. They verbalize that they are adopting communication patterns of younger people – patterns that do not come as naturally to them:

> If someone says there's a picture of so and so on Facebook, actually, we jump on and just check it out. We take what we get, whatever is there is what we enjoy. We have found with Facebook though it's been a huge difference in our lives because we do not get information directly even from our children. They forget to tell us because they assume that we have read it on Facebook. But we take what we can get – it's the young's way of doing things. We see this technology thing through old eyes. (Mary)

For Mary, Facebook is important because it helps her to maintain a connection to her children (and, therefore, her grandchildren). Socializers, like Mary and Gwen, often verbalize that they have adopted these ICTs (texting, social media, etc.) because their intergenerational relationships necessitate it. They do not expect young people to adapt to the usage patterns with which they are most comfortable. Instead, older adult Socializers believe that they will create a stronger relationship with young people if their own (older) generation is the one that adapts. Mary enjoys using Facebook, and what she gets from her use of it, because she values her relationships with her grandchildren and children.

Socializers prefer ICTs which allow connections to be made and maintained. This focus on using ICTs to strengthen relationships' permeates across all areas of Socializers' lives and across many different technologies, as Gwen speaks about:

> My cell phones I must have. That's a must. I must stay in touch with my family and my neighbors. So, my digital camera of course would be just for things happening here in the community and with neighbors, with the children. Sometimes not so nice things that happened in the neighborhood when the bad guys were out there, and I take their picture. Sometimes I need to call neighbors and they have to call me or they need to use the cell phone, or I need to call on my community for help. If there's for instance a birthday that one of us is put on the morning news, or someone in the neighborhood, a school age kid or a parent and there's an event, I watch it on TV. Or if someone has been in an accident and we want to hear about it on the TV. The focus is in what is happening in the community. Microsoft Word I use with writing. I send out poetry to people in the community. Facebook I use to just send messages and having fun in the community and among my family. I text for documentation, mainly different things and dates. Making sure dates are right when something bad happens and getting people's name and events, things that have happened for record keeping. (Gwen)

Gwen lived in Section 8 Housing (private low-cost housing with government provided rent assistance) and was often referred to as "Grandma Gwen" by the children living in the complex. She watched out for her neighbors, often speaking about the occasional issues that happened: domestic violence incidents, drug abuse, and criminal activity. In these cases, she used her ICTs, such as her digital camera and texting, to document occurrences, helping to protect her neighbors and build community. She ran a small food pantry out of her hall closet with help from her church, using her cell phone as a means of connecting to those who would otherwise go hungry. Every technology she has in her life, be it word processing, social media, or the television, is used in some way to strengthen relationships.

Mary was also deeply involved with charity work through her religious community and describes how she uses various ICTs to keep her connected to others:

> I'm on the telephone with shut-ins and I email our godchild. We're very pro-life and we're very concerned with all the issues that are going on. I'm always on the computer getting information on all kinds of activities. I have what I call a black book which I've had for probably since 1999, since I got the computer.

> It's full of things that's developed in the news and it was on all issues of concern that I had. One of the first things I out in there was when Dolly was cloned. It was in the newspaper and it was this little article and I cut it out and I showed everybody [...] I often share things I've kept in my black book with the members of the community. (Mary)

Socializers use all the ICTs at their disposal to promote, build, and maintain their relationships, even ones that many people would not consider communicative. Mary speaks about how she uses the Internet to research issues she is passionate about, from her pro-life stance to her views on cloning. The purpose of researching these issues, however, is not for her own benefit, but so that she can share them with others in her faith community and among family and friends. Her use of the Internet for research is not a task (as it is for Practicalists) or for fun (as it is for Enthusiasts) but is meant to *build community*.

Socializers have a unique ability (compared to other user types) to use ICTs that many would consider non-communicative or non-social to build relationships. Socializers view all ICTs as being potential connectors (if used correctly), even when other user types would not. Gwen used non-communicative ICTs for relationship building, including digital photography:

> My digital camera, I just loved it because I could capture, capture, capture. I could record. I love to capture moment by moment as the storm comes, the sky changes. I would keep the date on and the time. I could see from seconds, just really seconds to minutes, how things changed. It was just wonderful, just so wonderful. I sent my daughter and her husband a card that I made for their anniversary back in April. She said "Mom, this picture that you took was exactly a year ago on our anniversary!" So, the date was there and the time and everything, so it captures everything. It lets me be connected to people in a way I otherwise couldn't. (Gwen)

Most individuals (who are not Socializers) would not view or use their camera in the same way as Gwen. She does not use it just as a device to take pictures, but as a device to take pictures *to share with others*. In particular, this sharing focuses on relationship building. The card Gwen makes for her daughter is not made from any picture that her daughter might like, but a picture was purposefully taken at a specific date important to her daughter. The picture was not taken because it was pretty, it was taken because it could *build a connection*.

It is not the technology itself that necessarily encourages socialization, but Socializers' unique use of it that encourages such connectivity. Nancy, who lived in an assisted living center, was limited in the number of ICTs she had access to by finances and due to her impairments (arthritis and visual). However, despite having only limited access, she made use of the technologies she had to maintain her relationships, most notably her television:

> I like certain things on the television about India and they have specials on every once in a while. I like the specials that they have. My daughter was calling me and telling me about something that she had seen on the History Channel. Lots of times I'll talk and share what I've seen with my daughter Bette and my friend Danielle. I try to get people to watch TV together in the main recreation room. Danielle and her family donated that TV. I also find out about what my community needs on the TV [...] I had seen it on TV that they had a lot of things for needy children for Christmas on the news. But they didn't have anything for young girls, so I had the chance to make a lot of hand crocheted pocketbooks. I made a small purse to go inside and I think there was about 40 or some odd and I sent them in and they had them on the news. I will be doing that again. I saw in the newspaper last year that they needed helmet liners for the troops in Iraq. So, I told another lady here and she was helping me, and we sent close to 300 liners overseas. (Nancy)

Nancy used the television for three social purposes: to connect her with her friends and family by sharing what she has seen on television, to gather residents together to build community, and to find out about various volunteer opportunities. Nancy's television use reflects Socializers' ability to build a community with technology. Many people would likely believe the television is an isolating device, watched in solitude or used in place of a face-to-face conversation. The value in any ICT to Socializers is based upon their own personal views on how they can use the technology socially.

In contrast to Nancy, other Socializers expressed that they had little use for the television – except on the rare occasion when they used it for a social purpose. Gwen found little use for her TV, unless she had her grandchildren over:

> I'm not much of a TV watcher. Sometimes I think I have the TV on maybe just for the noise or I think you're supposed to turn the TV on once in a while or whatever. I don't sit still long enough to watch TV. So, I like to use it for the grandkids. So, when they come [...] I'm not interested in that stuff by myself. When they first came out with the TV I remember the living room was like the family place, you know, so nobody was shut off from one another. I mean mom and dad and the rest of the family watched the same thing. It was social. Kids today they have TV's in their own room and whatever. I would never have a TV in a bedroom. (Gwen)

Gwen's active lifestyle interferes with her ability to watch television. Television watching, she states, can be a social process (and in fact, she believes it used to be more social); however, in modern times it is often used in isolation. She has no use for television unless it is being used socially.

It is notable that Gwen had greater mobility than Nancy. Gwen also had greater access to many more ICTs she could use for communication, including a television, landline, computer, the Internet, and a cell phone. Nancy only had access to a television and a landline telephone due to her physical limitations and limited funds. This disparity in access can explain the differing attitudes of Nancy and Gwen toward television. In Nancy's context of the assisted living center, television was often the *only* available ICT many residents could use due to their physical, mental, or financial constraints. Given this context and her restrictions, Nancy prevailed in using the technology she had to foster community:

> We have televisions on all the floors. If somebody is in their room too much, getting too isolated or depressed, I try to get them to come to things we do here that other people like. Bingo is a big thing, and the Wii [gaming system] is a big thing. They have a bowling tournament on the Wii. I like to get other people involved in it. I like for other people to get involved in it. (Nancy)

Nancy used the television and digital gaming as two of the tools in her arsenal to promote socialization among the residents. She had served for years as the residents' elected representative at the assisted living center and had recently declined being re-nominated. Despite this change in her role, she still watched out for the other residents. She commented that many new residents had difficulty adjusting to living in the facility, and she sought to ease their transition into their new home.

A Wii gaming system had been donated to the facility about two years prior to our interviews. Many of the residents enjoyed playing group games, with bowling being a favorite. Nancy frequently encouraged residents to set up virtual bowling tournaments and she would often recruit participants from their rooms to play. She was not an active gamer, but she attended and organized these games. Nancy and Gwen's differing perspectives on television and video games reflect that the importance of an ICT to Socializers is not embedded in the ICT itself, but whether Socializers can use the technology to connect with others and/or build community in their everyday lives.

Despite Socializers' desire to adopt the communication technologies of those around them, such adoption can be limited due to physical, cognitive, and/or financial barriers. Due to Nancy's financial situation, she qualified for a United States federal government provided telephone program, the Lifeline program. This program includes a free cell phone with limited free minutes and data, provided by the government to individuals who receive certain social welfare benefits (Federal Communications Commission, 2018). The models that are provided, however, are often not designed for an aging population. Several of the participants I spoke with during the study had Lifeline program phones (most commonly through Assurance), but their models were tiny, oftentimes smaller than my digital recorder. For Nancy, who had come to the assisted living center due to her arthritis, this was frustrating:

> I had one of those little [Assurance] phones and that thing drove me crazy. It's so small that it makes it very difficult to try and use it properly because you're always hitting the wrong buttons. I found that if you have arthritis or you have anything wrong with your hands it's almost an impossibility to use a cell phone. So, I wish someone would somewhere along the line think about the elderly and give them a bigger phone, something that they can use on a regular basis. It was too small. I sent it back to Assurance. (Nancy)

Nancy's experience with the Assurance cell phone illustrates that Socializers, like all the other types, are not necessarily free to use the devices they would like to because of physical, cognitive, or financial barriers. In the case of Nancy, her arthritis left her unable to manipulate the buttons despite desperately wanting to learn texting:

> Texting is something that all of the kids are doing right now. My oldest great-grandchild is 13 and she's texting all the time too. I might be a little slower at pushing the buttons, but I would be able to do it if I wasn't overdoing it and hitting two or three at a time. They have the big phones and the large phone numbers for people, but I don't think anyone has thought about keeping the cell phones from going too tiny. There are people that are in their 20's with arthritis and I bet they're having the same problems we elderly are. (Nancy)

Nancy wanted to be able to text her children and grandchildren, a point she brought up in every interview, several times. (Her friends and family that I interviewed also mentioned to me that she wanted to text.) As a Socializer, Nancy wanted to communicate with the youngest individuals in her network the way that they communicated: texting. However, her financial situation and impairments left her with no opportunity to adopt the communication patterns of her grandchildren.

Socializers value constant communication with their large intergenerational family, friend, and community networks. They tend to have busy lives and are often out in their communities. Their need to constantly stay in touch, coupled with their high levels of activity, means they have a strong preference for mobile ICTs.

ICT Display

The more mobile a technology, the more useful Socializers find it. ICTs tend not to have specific places in their homes, but instead move with Socializers. Socializers prefer the cell phone to the landline because it allows them to communicate while doing everyday tasks: staying up to the minute with their social networks. They prefer tablets and laptops to desktop computers because of their portability. They move their ICTs with them around their home and outside of it.

Socializers are so in the habit of taking their communication devices with them that such behavior is automatic. Both Gwen and Mary carried around their cell phones during our tour of their ICT displays in their home. Their involvement in their neighborhoods and communities results in little time to be "chained" (Gwen) to an ICT that is not mobile:

> The cell phone I can walk around with. I'm not confined to sitting. I'm not just sitting while talking on the phone. Instead, I can get a lot of things done. I can make my bed. I can do a lot of things, like hang out my clothes. I can do a lot of things, like go to my car. I spend so much time on the phone that if I couldn't get things done while on it I'd never have clean clothes or make my bed. (Gwen)

Gwen relates that she spends so much of her day communicating that without a cell phone, she would likely not complete simple daily tasks, such as laundry. Socializers shared that they often kept their cell phones with them at the table while eating or in the kitchen while cooking. The motivation for keeping these devices with them is much different from Enthusiasts: Socializers keep ready access to their devices to be available for communication with others, whereas Enthusiasts keep ready access to be able to play with their devices. During our interviews, Gwen kept her cell phone on the coffee table, easily in reach from her position on the couch, despite her having shut it off (Figure 3).

Gwen, who did not own a computer but accessed the computer at the local library (mostly for social media and word processing), had an interest in owning a tablet or a laptop, primarily because they were portable. She had once owned a desktop but had found its stationary nature to be too "sedentary" and incompatible with her lifestyle. Mary and her husband, Cleveland (a Practicalist) owned a laptop they shared. Mary had "finally convinced" her husband (over the course of our meetings) to leave the laptop in the living room (where she spent most of her time at home) rather than placing it in the basement office:

> I like to have the laptop up here, so I can just use it. We used to leave it downstairs, Cleveland likes to use it down there. But I don't see the point of leaving it down there. I like to check my email or Facebook when I feel like it and not having to go down to the basement. (Mary)

Figure 3. Gwen's Cell Phone Is Never Far From Her Reach.

Mary, as a Socializer, wanted to be able to use the laptop to communicate when she felt like it (which was quite often, she checked her social media and her email before and immediately following our visits together). Cleveland, as a Practicalist, viewed the laptop as a tool that belonged in a tool-specific space: the basement office. It was notable that during my interviews with Mary, Cleveland would often retreat to the basement office with the laptop. He would return at the end of the interviews with the laptop in hand to give to his wife, who would immediately check her email and social media accounts. If Cleveland failed to bring up the laptop from the office, Mary would gently remind him to do so. As a Socializer, Mary wanted her devices easily within reach, so that she could constantly be in touch with others.

Nancy was much more limited in her use of more advanced ICTs due to her arthritis and finances. She owned a flat screen television and a large button land-line telephone. Her room in the assisted living center, while private, was quite small, and therefore, all devices were within easy reach. During our interviews together, Nancy sat in the only chair in the room and I sat on her walker at the end of the bed. She always moved her phone closer to her during our interviews so that she could "answer it if anyone called" – which they frequently did. (Our interviews were often interrupted by five or six phone calls.) Much like Gwen's movement of her cell phone to the coffee table during our interviews, Nancy moved her telephone to always be close at hand.

Despite Nancy's inability to obtain a cell phone that she could manipulate and afford, she made her own mobile technology: a simple small notepad and a pen, which she stored on her walker in a crocheted pouch she had custom-designed and crafted (Figure 4).

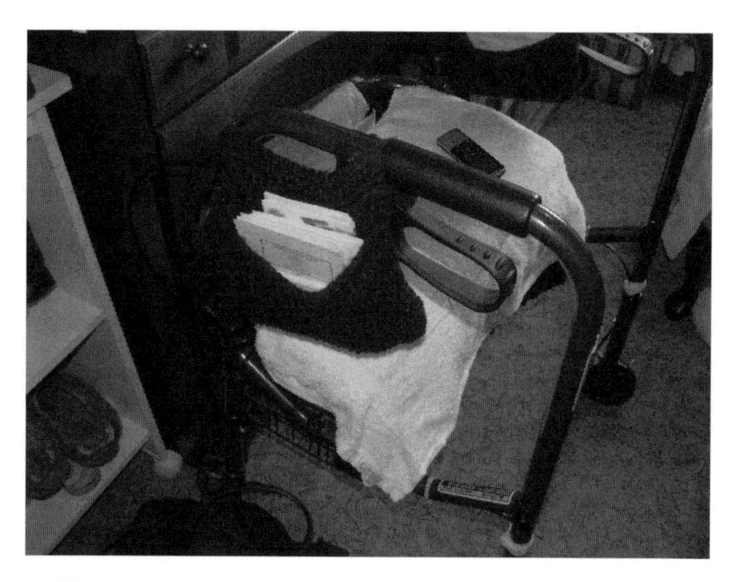

Figure 4. Nancy's Walker with her Portable Pad and Pen.

Nancy, as a self-appointed "advocate" who watched out for her fellow residents, often recorded incidents that occurred in the assisted living center in her notebook. These included events such as residents not receiving proper medications, staff misbehavior, or other incidents. (This documentation is similar to Gwen's cell phone and digital camera documentation of the "not so nice" events that she saw happen in her own community.) Nancy also took notes on things to relay to her fellow residents' families. During my visits to the assisted living center, I saw almost every walker outfitted with such a crocheted bag and many wheelchairs featured them as well. Nancy made these in custom colors and with customized storage for other residents, who appreciated not having to carry a separate bag. Nancy shared that she would keep a cell phone in her custom bag: "If I could get a cell phone I could use the buttons on, I would keep it on my walker. That way I could text or call people anytime."

For Socializers, the portability of their devices is key to allowing them to stay in constant contact. Carrying their devices with them (and quickly answering calls and replying to texts) sends the message to others that they are always available, and value communication and their relationships. Socializers welcome such constant contact.

ICT Meaning

Socializers' lives center on family activities, religious activities, volunteering and community involvement. Socializers are deeply embedded within their neighborhoods:

> I'm on the telephone with shut-ins and I email our godson. On the computer I'm always checking out things I am interested in and causes we're active in. We're very pro-life and involved with our church. We're very concerned with all these issues that are going on so I'm always on the computer getting information on all kinds of activities and things that are happening [...] My family is most important to me – we raised five children and they will always be the most important people in our lives. (Mary)

> We [the residents] wanted to get a flag for the assisted living center and I had written to our legislation and asked different politicians, and they finally got us a flag. Then we wanted to get a flag pole so we could have one outside and we worked to get that. We wanted to have a gazebo so we did all kinds of things to raise money to get that gazebo on our own. So, there's been quite a few things I guess I was involved with here. That's not even getting into my crocheting for charity and for the residents here. I run the community store. (Nancy)

> I think it's kind of natural for me to be involved in the community. I did not intend it to be so much – I've been honored quite a bit you know from different groups [...]. I just want people to respect themselves. I just want to help. I'm just Miss Gwen but Miss Gwen has a lot of fingers that reach and a big heart that reaches. (Gwen)

For Socializers, community involvement and family play a central role in their lives. Mary is involved deeply in her family and in her charity work. Nancy undertakes countless tasks to promote community among her fellow residents (from installing a flag pole to running a community store that provides everyday items residents would like at low cost). Gwen is involved in her church, runs a food pantry out of her closet, and is closely involved with her more than 20 grandchildren. The value of an ICT to these individuals is the connection the use of that technology allows. To Socializers, ICTs can deepen existing relationships and form new ones:

> Another thing that we really enjoy about the computer is getting pictures sent to us from the children. Our son has muscular dystrophy and he was home for Christmas. While he was home his cousin installed a chairlift at their house. It is wonderful that he has it. We wanted to see what it looks like so our daughter-in-law sent a picture of it over email. It was so tremendous to see how it fit in because of course we're visualizing it taking up the whole stairway and I thought that really was very nice to actually see it when we couldn't travel there. Then we gave a e-reader to our

> daughter and she sent a picture of our grandson who is 7 sitting there reading a book on the e-reader. So, I've been appreciating more the connection we get from the computer. (Mary)

Mary, like Nancy and Gwen, frequently used the word "connection." An ICT goes beyond the simple function of communicating to Socializers, to actually deepening and sustaining relationships. The word connection emphasizes the importance of technology in helping to create an emotional bond. This perspective is very different from task-focused Practicalists or fun-focused Enthusiasts. Socializers tend to love their technologies because of their potential to connect them to the people they love and care about.

Socializers' focus on wanting to use the ICTs that their younger friends, family, and community members use is not an attempt to "look young" or "hip." (Although most of their friends and family rate them as much younger than they are, credited in part to their willingness to text.) Instead, this focus on using, learning, and owning the ICTs of "the young" is due to Socializers' deep desire to be in touch with the younger generation.

Socializers place a high value on those ICTs which they view as deepening relationships and strengthening community and have little use (or value) for ICTs which they feel are isolating or intended to be used alone. It is important to note that this judgment of what "brings people together" versus what "separates and isolates" is in the eye of the individual Socializer and is deeply impacted by their context. Nancy, for instance, saw video games (her assisted living center owned a donated gaming console) as fostering community:

> I like the Wii. We'll have tournaments on the Wii, and people love it. Especially people who like to bowl but can't physically do it anymore. It really brings the community together in a way that we otherwise couldn't because so few of us can be active. (Nancy)

For Nancy and her fellow residents, virtual bowling was a fantastic community-building activity. Many residents participated, excited that despite their impairments and/or disabilities, they could bowl once again. The gaming console, used in this way, was a socialization technology. Many outside of the context of the assisted living center would likely characterize digital gaming as an *isolating* activity, particularly when compared to undergoing a similar activity in a physical (non-virtual) setting. However, because virtual bowling fostered community in her assisted living center, to Nancy, as a Socializer, it was a *social* technology, not an isolating one.

As stated before, Gwen had a different view on video games and television use. Gwen did not enjoy watching television, preferring to spend her time reading, writing, or socializing. She felt that the television was isolating and the only time she really watched it was when she had guests over to watch a program, particularly her grandchildren. She was not a fan of video games, believing them to be isolating. Both Nancy's and Gwen's very different opinions on the ability of television and digital gaming to connect them to others reflect the values Socializers' impart to technology: ICTs which are social are valued and heavily used; ICTs which isolate them and cause disconnection are not. While Nancy uses the television to connect to others, Gwen does not; so Nancy is a relatively heavy user of the television while Gwen rarely turns hers on. The meaning of the television is not in the television itself, but rather exists in the meanings the individual assigns the device (Lie & Sørensen, 1996; Silverstone et al., 1994). Socializers, like all the types, cannot be identified by the ICTs they own or use; but by the meanings they ascribe to these ICTs.

Both Gwen and Mary were writers, and they viewed the computer and word processing (upon which they both were writing books) was a way for them to share their faith and beliefs:

> I do a lot of writing [...] It's important to me that I write every day I can. Writing is so important. It has helped me heal from the trauma in my life. Writing has helped me heal and has helped me share. Family is just so strong, so strong and so important. Passing history on is very, very important to me. Writing is not only important to me and my family but just passing things on is really important to do. So much of our history [as African Americans] is lost, and it's so important to be passed on. There's a proverb and it says if you have knowledge let others light a candle to it and I certainly believe that is important – that I share my knowledge with others. (Gwen)

> Now I probably would not have written a book if I didn't have Word, because I do not have a good handwriting. I would just scribble it all out. But when we got the computer and Word it was absolutely a huge gift. I really am a big admirer of Word because it has made writing a book so easy to share and write. My book is very important to me, it's about my faith and life as a mother. (Mary)

For both Gwen and Mary, similar themes of the importance of family emerge as they speak about the importance of writing, and the role of word processing in enabling the sharing of their thoughts with others. For Gwen, it is the sharing of a previously mostly oral African American cultural history and tradition. For Mary, it is the sharing of her faith and family values. Their writing is not a task to be completed on a simple tool, but rather an act of *sharing* – a word both women used independently to describe their writing. Word processing is seen as an enabler of this sharing, of building community and family. As spoken about previously, Gwen used her digital camera as a device to build relationships, using printed photos to make greeting and thank you cards. She felt this let her personalize her cards, making them more heartfelt, drawing her closer to their recipients:

> I love photography. I think the skies and the trees tell us such great stories. I make cards out of some of my photographs and I send people out different pictures I take. I write a little poem that goes with the picture. My soul speaks to me and I just write out what comes into my head and pick out a picture that is special to me. It is made for that specific person, made only for them. That makes it special. I love my digital camera, because it allows me to make these special cards. (Gwen)

For Gwen, the value in her digital camera is not that it is a fun toy to play with (as it would be for Enthusiasts) or a great hobby tool to use during leisure time (as it is for Practicalists), but the value lies instead in making a personal and "special" connection to others. She loves her digital camera because it deepens her connection with those she creates these cards for. Creating these cards is both deeply personal and deeply social; it is an act of love. For Socializers, ICTs are valued for enabling, facilitating, and strengthening relationships.

Socializers: The Technological Social Butterflies

Socializers are well connected with their communities, families, and friends; and are often involved in charity, volunteering, and/or their religious communities. This high level of involvement means that they have a large network of intergenerational contacts. They spend a large part of their time socializing and communicating with others. Technology enables such socialization and they can be seen as the "technological social butterflies" of the ICT user types. Key points about Socializers include:

- Socializers deeply value relationships, building community, and sharing knowledge.
- Technologies used for socialization activities are valued, those they see as non-social or isolating, are not.
- In elder age, Socializers are eager to learn the technologies that young people are using, mimicking young generations' use patterns to foster intergenerational relationships.
- Socializers prefer mobile technologies that allow them to be in constant contact with their large intergenerational social networks.
- To appeal to Socializers, one should emphasize the social nature of a technology, or its ability to connect individuals and deepen relationships.

While Enthusiasts love ICTs as fun toys, Practicalists see ICTs as tools, and Socializers value ICTs for their potential to connect them to others, the fourth user type, the Traditionalist, values the ICTs of their youth. We will explore Traditionalists and their nostalgia for these older forms of technology in Chapter 5.

Chapter 5

Traditionalists: The Keepers of Technological Tradition

> I'm not against future things. Like my son said to me over the weekend, he said "how's your computer coming?" I said, "It's alright." He said, "Mom you got to really pay attention." I said "Thomas, if I'm going to call somebody I'm going to communicate with someone, I would prefer to call them on the phone and listen to their voice." Sometimes you can pick up things quicker than if you're going to type it out and then wait for a response, at least you'd be there in time if your friend needed you right away. I have a friend with depression. If she sounds depressed, I can say "Let's go for a walk. Let's get out of the house." You don't get that from an email. Email is good for some stuff, not for me. My son doesn't get that. How can you not get that? (Mindy Jean)

Nostalgia. Tradition. Memories. Love.

Speaking to Traditionalists is to hear a love sonnet for the "traditional" ICTs of their youth. Notably missing from these verses, however, is any mention of ICTs introduced after Traditionalists reached middle age. In the case of the Lucky Few Generation, they wax poetically about the telephone, television, and radio while seeing little use for cell phones, computers, or the internet.

Traditionalists come from a variety of work backgrounds. Some have exposure to ICTs in their work, while others have none. They heavily use the ICTs of their youth and heartily reject more modern forms of ICTs. They fill their homes with traditional technologies, while hiding more modern ones they often receive as gifts from well-meaning family and friends. Overall, Traditionalists express a great love and nostalgia for the technology of their youth – so much love, in fact, that they find their lives too full of these ICTs to have any space for more modern forms.

Formative Experiences

Traditionalists often express warm memories from childhood about their use of ICTs. For the Lucky Few Traditionalists in this study, this included warm childhood memories of the radio and (when they were young adults) television. Mindy Jean, who had originally chosen to stay at home with her children, later continued to stay at home to take care of her aging mother. She was a dedicated soap opera fan, having listened to them since she was a teenager:

> I'm a soap opera fan. I used to watch a lot of soap operas. I used to listen to them on the radio more so than anything, and some of them got on TV. Golly, I bet I've been watching them at least 50 years, at least. I listened to them as a teenager [...] I used to listen to soap operas on the radio when I was home sick from school. I'd listen to all the soap operas with my mother. "*Guiding Light*" I used to listen to, "*As the World Turns*" I think that was on [the radio then], there was another one, I can't remember what the other one was. Oh, I just listened to them all, I remember listening to them [...] It was great [when they moved to television], you could see who was talking, the voice of the person, you know, what they looked like [...] It was sad [when they were cancelled]. I watch one now from 12:30 until 1:30, "*The Young and the Restless.*"

Mindy Jean shares that she began listening to soap operas on the radio with her mother when homesick. She became a lifelong fan, transitioning her soap opera consumption from the radio to the television, and listened/watched some soap operas from their near beginning (such as "*Guiding Light*") to their very end.

Traditionalists tend to share feelings of warm nostalgia over their use of ICTs in their childhood and young adulthood. They have a strong preference for the programs (particularly television programs) of their youth, although they often enjoy other newer programs as June shares:

> I look at a lot of the shoot'em ups [Westerns] from when I was younger. Which I love, and I look at lot of reality television. We know reality TV is lies, but they can be very interesting; the housewives fighting each other and all that stuff. I ain't never seen anything so silly. (June)

Lucky Few Traditionalists, like Mindy Jean and June, enjoy the nostalgia that listening to "oldies" music and watching rerun television programs from the 1950s through the 1970s evokes. Unlike Enthusiasts, they have no memories of tinkering or being encouraged by technology mentors. Instead, the warm nostalgia of Traditionalists often focuses on the experiences of sharing ICTs with their family members. While Traditionalists have many avenues through which they are introduced to new ICTs, including work and family members; their default preferences are always for the ICTs of their youth.

Introduction to ICTs

Traditionalists have diverse background experiences with ICTs. Some were first exposed to computerized ICTs through work, while others were first exposed through their families. Traditionalists are not morally opposed to more modern forms of ICTs, so they will often try new technologies introduced by family members:

> One of my sons is going to come up and set up email for me. I'll try anything, so they come up and set it up. I'll try it with them. It's not anything I really desire. I'm not fighting for it. I'll try it – that's it. (June)

> I'll try the computer. But I'm not really that interested in it. (Mindy Jean)

As both June and Mindy Jean express, Traditionalists are willing to try new forms of ICTs, be it email (for June) or using a laptop computer (for Mindy Jean). However, their use does not become an established practice. These devices come to occupy a desk drawer or gather dust once Traditionalists determine the trial period is over (which tends to occur quite quickly). Traditionalists often find that their disinterest in using more modern ICTs is met with concern by their families, particularly their children. These children tend to buy them all sorts of devices, concerned that Traditionalists are "out of touch" or falling behind. June had received several gifted cell phones:

> My daughter bought me a cell phone when they first came out [...] Then I got one of these that we call the Obama phone, an Assurance phone [...] My girl friend moved to Texas and she gave me this other phone because she said she wanted me to forever keep in contact with her. But I don't use them. I use my landline. (June)

June had received two cell phones as gifts, one from a friend and one from her daughter. She also had a government-provided cell phone. Despite having multiple cell phones and the encouragement of her friends and family to use them, June continued to use her landline phone. This behavior typifies Traditionalists. While most would assume Traditionalists lacked access to advanced/ digital ICTs, such was not the case. In fact, Traditionalists in this study tended to own as many, if not more advanced ICTs than some individuals of other types, including some Socializers and Practicalists. All Traditionalists owned cell phones and computers that had often been given as gifts to encourage their use:

> I have a notebook computer I got last year for Christmas from George [my husband]. My son set me up and he said he was

> going to help me learn how to use it. That never happened. My daughter has helped me a little bit and gave me some directions on how to use it and then when I did the directions something else appeared on the screen. I could never get past that point and then when she explained it she went so fast. I needed a little more time. (Mindy Jean)

Mindy Jean had received her computer from her husband as a gift and was often encouraged by her children to use it. However, she found their directions and lessons confusing. At first glance, it might seem that with more assistance Mindy Jean would be a successful computer user. Such a thought suggests that her difficulties in learning were due to a lack of knowledge. However, she often shared that she simply was not interested:

> I'll try the computer. But I'm not really that interested in it. (Mindy Jean)

At first, in our conversations, Mindy Jean was reluctant to discuss her disinterest. As our time together went on, however, she began to share that she really had no interest in using the computer, but did so just to please her husband and children. Traditionalists do not necessarily reject more modern ICTs out of lack of knowledge, but rather out of such disinterest. (June, for instance, had been a legal administrative assistant/ paralegal prior to retirement and had worked extensively with computers in that role.)

Compare Traditionalists, who have low levels of motivation, to individuals of other user types who are highly motivated, but lack access to more modern ICTs or knowledge of them, such as Nancy (Socializer) and Dan (Practicalist). Nancy, a Socializer, expressed many times that she wished to learn to text, but she was unable to find a cell phone that she could afford and that would accommodate her physical impairments. Dan, a Practicalist, faced challenges in learning to use the computer. These were due, in part, to having worked in a position that did not require computer use, coupled with the lack of technical training and support after retirement. Both Nancy and Dan expressed strongly *wanting* to use these devices and programs and they went to great lengths to do so. Nancy tried multiple cell phone models. Dan sought out computer lessons from his wife. While Mindy Jean would speak about how she would try an ICT, she never expressed the same level of motivation to learn as Nancy or Dan despite having many ICTs available:

> I'm not against future things. Like my son said to me over the weekend, he said "how's your computer coming?" I said, "It's alright." He said, "Mom you got to really pay attention." I said "Thomas, if I'm going to call somebody I'm going to communicate with someone, I would prefer to call them on the phone and listen to their voice." Sometimes you can pick up things quicker

> than if you're going to type it out and then wait for a response, at least you'd be there in time if your friend needed you right away. I have a friend with depression. If she sounds depressed, I can say "Let's go for a walk. Let's get out of the house." You don't get that from an email. Email is good for some stuff, not for me. My son doesn't get that. How can you not get that? (Mindy Jean)

As Mindy Jean shares, she is not interested in using email; however, she often faces pressure from her loved ones to do so. She shares that her son, Thomas, scolds her to "pay attention" and start using her computer; admonishing her as if she were a child. Traditionalists often feel pressure from family members to use ICTs, particularly ICTs that were given as gifts. Traditionalists often feel guilty that they have been given these expensive gifts and will try the ICTs. But their lack of interest means their use is not sustained.

As a result of Traditionalists' high consumption of more traditional forms of media such as television and newspapers, they are often very aware of new ICTs. This knowledge tends to be very basic and they tend to not understand how a specific technology works in detail. However, Traditionalists do understand that these devices and applications exist. All Traditionalists spoke of various forms of social media, including Twitter and Facebook. Traditionalists often watch, hear, or read about these innovations on the traditional media forms they love:

> If you say something to a friend in confidence and you say that's between you and I and your friend keeps it between you both. If you put it down on in a computer and email or Facebook it's not always between you and I anymore. It can leak out in ways I guess – it gets passed on or forwarded and sometimes this is even by mistake. They had that on the TV the other night. And there's bullying that happens. That's why I don't want Facebook, I don't want to miss the actual interaction with my kids and grandkids and I don't want to participate in something that can hurt people. I know that other people love it and use it, but it's not for me. (Mindy Jean)

Oftentimes, the knowledge that Traditionalists have of more modern ICTs simply informs them that they are not interested in using them or any of their potential features. For instance, Mindy Jean expresses how emails and social media can be forwarded on and conversations are not private. She has access to more traditional forms of ICTs which allow for, in her opinion, richer communication. Such traditional technologies are the core of Traditionalists' use.

ICT Use

Traditionalists have a strong preference for doing technological things "the old way." In the case of the Lucky Few birth cohort, this means choosing to use the landline over a cell phone, the TV over the internet, and the radio over a digital music player.

Traditionalists believe that the ICTs of their youth are far superior to more modern ICTs which have been developed as they have grown older. They are extremely heavy users of these "traditional" ICTs and use them in every context of their lives. Mindy Jean shares how she is continually consuming media in her home:

> I love my radio. I like soft rock. I like the older tunes too. I like the up-to-date songs that some of them play. I don't like any of the rap or anything else like that, that's not my bag. CD players, or my player, I play that, I play my discs. I just like music. I used to listen to soap operas on the radio as a kid when I was home sick from school I'd listen to all the soap operas. "*Guiding Light*" I used to listen to, "*As the World Turns*" I think that was on, there was another one, I can't remember what the other one was. Oh, I just listened to them all, I remember listening to them. I used to watch them on television too – until they got canceled. Now I watch "*The Young and the Restless.*" I didn't always sit down and watch TV, my soap operas, like just sit there and constantly watch. When I was taking care of children in my home and they were napping, I would sit down that hour and watch that soap opera because it would give me a chance to get refortified, 'cause I had busy kids [...] I mean I wasn't a fanatic, but they came on at different times that I had a chance to sit down or if I was busy in the kitchen I'd have the TV on [...] if I don't have the TV on I have the radio on. I play the radio when I have friends over, when I do housework, or really anything. (Mindy Jean)

Be it the television or radio, one or the other is always on at Mindy Jean's. Even when she entertains, she often has the radio playing in the background. It was not uncommon for Traditionalists to be using one of their ICTs when I met them or called to arrange for a meeting, or even in our interviews themselves. June commonly left her television on for background noise during our interviews together. Such constant use is similar to Enthusiasts, who also tend to use ICTs in every facet of their lives and use them heavily. However, while Enthusiasts love all ICTs, with a preference for newer forms, Traditionalists love only the forms of their youth.

Whereas Mindy Jean speaks fondly about using the television and radio, she shares that her use of the cell phone is limited:

> I'm not that interested in using my cell phone, let's put it that way. Right now I'm able to communicate and find things out my normal way by using the landline. When I can't do that anymore and then I'm really going to have to check things out a little

> more seriously. But I don't know if that's going to come in my day and age. Probably I will need to learn. My kids had to call me when I got home in the old days; I never had an answering machine. If a person wanted you enough they'd keep calling until they got you. Now with the cell phone you can get in contact with them right away. Well I wasn't too interested in carrying a cell phone and I don't necessarily always have it on and that drives certain people crazy. Because like I said it's for my use. I really don't want everybody calling me on it. Only my family has my cell phone number. They all think I'm a little crazy, but I already have a phone. (Mindy Jean)

Mindy Jean chose to only give her cell phone number out to close family, uninterested in using it in day-to-day contexts. This non-use stands in contrast to Mindy Jeans' use of the traditional form of the landline telephone, which was very extensive:

> I used to call my mom every day when she wasn't living here. I call my daughter pretty much every day or twice a day. I like to chat on the phone. I have a few people I like to call [...] The land line is the phone number that we normally give out to anybody. The cell phone numbers we kind of keep to ourselves. You don't want everybody in the world to know your cell phone. Our land line that's the phone that we give out on applications or in doctor's offices and so on, so we do get reminded of our doctor appointments and stuff like that. We use the landline to call out most of the time for doctor appointments or we use it for our friends. I'm more in contact with friends through the landline than my cell. My cell is for emergencies. (Mindy Jean)

This juxtaposition in Mindy Jean's use of the cell phone versus the landline is typical of Traditionalists. This user type has a very limited use for the ICTs that were developed after they reached middle adulthood. For the Lucky Few generation, rejected ICTs include cell phones, computers, and social media. Embraced ICTs include landline telephones, newspapers, magazines, books, television, and radio.

It is important to note that many Traditionalists have access to, or even directly own, the ICTs they refuse to use due to gifting. For instance, June had a Facebook account that had been set up by one of her daughters. At June's request, her daughter only friended family on June's account (as opposed to friending friends or former colleagues, etc.). June only visited Facebook when one of her children encouraged her to do so:

> My family sends me a message. They'll call me and say, "Go on Facebook. I just put something on there for you" and that's

> when I go in and look for the message. I only use it for family. And I only go there when I'm told that there is something there for me. I don't get why they just don't tell me what it is on the phone. I mean, they're already talking to me! (June)

Traditionalists severely constrain how many life contexts they use modern forms of ICTs in. For June, Facebook was an application she only used with family and she only used it when prompted (somewhat begrudgingly). In contrast to June's use of the television, which she used for leisure and with family and friends, her use of Facebook was targeted. This is similar to how Mindy Jean frequently chatted on the phone, but only used her cell phone in emergencies, and only her immediate family had her cell phone number. Such strict limits on modern ICTs stand in contrast to Traditionalists' heavy use of older ICT forms.

As more information and services have moved online, Traditionalists have adapted. While not direct users of advanced forms of ICTs beyond "trying" them a few times, they are often "indirect" users of advanced ICT forms. They do not directly manipulate the ICT, but, instead, they have others that directly use such devices to get the specific information they need. To facilitate this indirect use, Traditionalists tend to develop a network of individuals who are willing to complete internet tasks for them:

> If I really need something off the internet, someone will find it for me. That's why I have kids. And if my kids won't do it there's a man who works at the front desk who's into this whole technology thing and he'll do it for me. (June)

Traditionalists often rely heavily on direct users for digital tasks, just as June relies on her children, or the man who works at the front desk of her low-income apartment complex. Oftentimes, there is concern in the literature that non-computer users will be "victims" of the digital divide – unable to access important information or services on the internet unless they learn to use computers themselves (Paul & Stegbauer, 2005; Van Dijk, 2005). Such concerns fail to recognize that Traditionalists are often already getting the services or information they need by being indirect users. Mindy Jean and her husband share how they use the computer in their relationship, with Mindy Jean (a Traditionalist) being an indirect user while George (a Guardian) is a direct user:

> The computer is really my husband's thing. If I really needed to do something online, he could do it for me. (Mindy Jean)

> I see the computer as my thing – as my responsibility. Using the computer is my duty. If Mindy Jean needs something, I do it for her. Online shopping, finding out the weather, all those things

I can do for her, so she doesn't need to bother trying. (George on Mindy Jean's computer use)

One easily senses that George believes that Mindy Jean is quite capable of using more modern ICTs, if she would only "bother trying." Traditionalists are often met with such judgments on their non-use by family members.

Being non-users of more modern ICTs does place Traditionalists at risk of being excluded from our increasingly digital societies, particularly if the direct user they rely on can no longer acquire information or access services. (Meeting the needs of Traditionalists in such situations is addressed in Chapter 10, which focuses on the practical application of the ICT User Typology.) Despite their non-use of more modern ICTs, many Traditionalists own them, resulting in unique ways in which they display these unwanted ICTs in their homes.

ICT Display

Walking into Traditionalists' homes is like walking into technological time capsules: one immediately sights many traditional forms of ICTs. These ICTs are displayed prominently and are the centerpiece of many of the rooms, as the digital hearth (Flynn, 2003). Notably missing, however, are technologies developed after the Traditionalist reached middle age.

Mindy Jean had four televisions in her home, located in every living place she commonly spent time or expected her guests to (the living room, kitchen, master bedroom, and guest bedroom). Each of these televisions was the focal point upon entering the room, as can be seen in her living room (Figure 5).

Traditionalists often own many duplicates of their favorite ICTs in order to facilitate their constant use, as Mindy Jean does. Mindy Jean's placement of

Figure 5. Mindy Jean's Living Room Television.

these televisions throughout her home meant that as she completed other tasks or moved about, she could constantly watch or listen to the television. She placed other ICTs she used frequently in locations where they also could be continually used. Radios were placed in the kitchen, living room, and master bedroom.

While Mindy Jean loved soap operas, her husband, George (a Guardian), was often critical of her watching "mindless television" (George). After George semi-retired (from full to part-time work), Mindy Jean often watched her programs on the master bedroom television when he was home. During days when George was working, Mindy Jean watched her soap operas on the kitchen or living room televisions. She often preferred to watch these shows in the kitchen in order to prep meals and complete craft projects throughout the day, making her time more productive:

> I used to watch the soap operas while I used to prepare dinner. "*Guiding Light*" used to come in at a good time, because I'd have it in the background while making supper. "*The Young and the Restless*" was at lunch time 'cause I'd sit here and eat my lunch and watch. (Mindy Jean)

Mindy Jean's watching habits had changed significantly since several soap operas had been canceled. Instead of listening and watching the television during dinner preparation, she had begun listening to the radio, another favorite technology.

June had three televisions in her home, all were located in the living room. She would have preferred to have a television in the bedroom but was unable to move one of the large televisions into that room by herself. The televisions occupied much of the living space and they were centered in the room.

Both Mindy Jean's and June's placement of televisions in their respective homes suggest the importance of this ICT to their lives: it is fundamental. Technologies which are valued hold prominent places in Traditionalists' homes and they own many duplicates of these. For Traditionalists, it is the form of an ICT that is important to them (for instance, radios and television were preferred to computers by Lucky Few Traditionalists). It is not the vintage of the specific items they own that carries weight. Having a 1950s or 1970s television that mimics furniture is not considered important to Lucky Few Traditionalists. Instead, having a television (of any vintage) is important. Traditionalists evoke nostalgia through their experience of using the technology, not necessarily the way it looks.

Traditionalists often have an "if it isn't broke, why fix it?" (Mindy Jean) mindset, however. This means that if a device is still usable, they do not update to the latest version or technological development. One will often encounter older devices in Traditionalists' homes. Their televisions may be cathode ray tube (CRT) televisions as opposed to liquid-crystal display (LCD) flat screen versions. Rotary dial phones are often still present (particularly in lesser used rooms). This is in contrast to many Enthusiasts and Practicalists, who tend to update frequently. Enthusiasts update due to the excitement of having more modern ICTs; Practicalists update due to new features. Traditionalists, on the other hand, only update when necessary.

Traditionalists tend to relegate their newer forms of ICTs that have been gifted to them to a spare space or pack them away. These are hidden in contrast to their proud display of more traditional ICT forms. June placed her computer in a corner of her main living room. (She would have preferred it to be in a spare room, but due to the space limitations in her small one-bedroom apartment, it needed to be placed in her main living area.) June's computer was not visible unless an individual entered her living room, turned 180° around, and sat in one of two positions on the sofa. It was otherwise blocked from view by a large over-stuffed loveseat.

Mindy Jean's husband, George, was a computer user, and she had him place his computer in an office in which she never entered, except to clean. When I asked George why his computer was in the office, he stated: "my wife decides where things go in the house." As a result, the computer was out of Mindy Jean's sight and did not interfere with her own ICT use: George could be using the computer while Mindy Jean watched television. Mindy Jean also concealed the laptop her husband had purchased her as a Christmas gift, placing it (along with her cell phone) in a locked cabinet in the living room (Figure 6). She shared that this was due to laptops not being appropriate for living rooms:

> I keep my notebook computer in the desk. I don't think it needs to be out all of the time. It's not really nice enough to have out in the open. I make my husband, George, keep his computer in the office. I wouldn't let him have it out here in the living room. Computers don't belong in the living room. You might have a TV in the living room, but not a computer. (Mindy Jean)

Newer devices, which are often pushed on Traditionalists' by concerned family members, are placed in drawers, under the bed, or behind the sofa; out of sight and out of mind. This split in what is displayed versus hidden reflects the meanings ICTs hold for Traditionalists: older forms are valued; newer forms are rejected.

ICT Meaning

Traditionalists denote strong attachments toward the older ICTs of their youth but show indifference toward newer forms of ICTs. They describe their relationship with older ICT forms as "love:"

> I love my television. I watch it all the time. I really love the TV. I like the older shows [...] (June)

Figure 6. Mindy Jean Keeps her Digital Camera and Computer Hidden in this Desk.

It is common to hear Traditionalists speak about their ICTs in loving and endearing tones. However, Traditionalists differentiate between ICTs based on their relative age. The important time marker in this differentiation is when a Traditionalist is middle-aged: technologies introduced before this time period are loved, those introduced after are rejected.

Traditionalists use the ICTs they love, much like Enthusiasts use the ICTs they love: during all waking hours. Traditionalists have a warm relationship with the ICTs of their youth and tend to speak about these ICTs as more of companions than as devices or media:

> "*Guiding Light*" I used to watch once in a while. That was on during my time that I used to prepare supper for everybody to come home. I'd have it in the background. "*The Young and the Restless*" was at lunch time and I'd sit here and eat my lunch and watch. I used to watch "*As the World Turns*" and that went off the air too. I remember "*The Guiding Light*" and "*As the World Turns*" used to be on the radio. They came on at different times that I had a chance to sit down or if I was busy in the kitchen. It was nice to have something playing in the background; otherwise you get lonely all day. (Mindy Jean)

Listening to Mindy Jean speak about her experience with soap operas reminds one less of a person speaking of a tool (such as Practicalists would) and more of a person speaking of a close friend. Her television watching provided companionship, particularly during her days of caretaking, first of her children and later of her mother. It provided respite and relaxation. For Traditionalists, media and the devices used to consume them go beyond simply offering short-term companionship to representing lifelong relationships. Using such media and devices allows them to re-experience their lives and, most importantly, relive the bonds they have for the important people in them.

Mindy Jean had begun listening to soap operas on the radio as a girl, when she was home sick from school, as her mother listened to them on the radio at the time. When Mindy Jean became a mother herself, she began watching the soap operas on television (as they had moved to that form of media). Soap operas became an important bonding experience between these two women, who would talk about the characters and plot lines unfolding. This discussion first started when Mindy Jean was a teen, continuing when she moved out on her own, and carried on to when her mother moved in with her during the final years of her life:

> I'm a soap opera fan. I used to watch a lot of soap operas. I used to listen to them on the radio more so than anything, and some of them got on TV. Golly, I bet I've been watching them at least 50 years, at least. I listened to them as a teenager […] I used to listen to soap operas on the radio when I was home sick from school. I'd listen to all the soap operas with my mother. "*Guiding Light*" I used to listen to, "*As the World Turns*" I think that was on [the radio then], there was another one, I can't remember what the other one was. Oh, I just listened to them all, I remember listening to them […] It was great [when they moved to television], you could see who was talking, the voice of the person, you know, what they looked like […] It was sad [when they were cancelled]. I watch one now from 12:30 until 1:30, "*The Young and the Restless.*"

The warm nostalgia Mindy Jean feels toward soap operas is not simply a love of this medium or type of show, but rather is a reflection of the love she had for her mother and for their shared connection and interest. These shows, for her, came to represent her relationship with her mother and she felt connected to her mother (even after her mother passed away) when she was watching them. Losing these shows was not merely the loss of entertainment, but also the loss of a connection to her mother.

Traditionalists do not tend to have negative feelings toward newer ICT forms and, in many cases, are not resistant to trying them. However, upon exploring them, they simply cannot see "what the fuss is about" (June). Their attitude toward the new ICTs is one of indifference (not negativity). During the course of our interviews together, Mindy Jean asked me to help her explore some online shopping sites. Upon viewing several sites, she thought would have excellent deals, she was disappointed:

> Everybody talks and talks about the computer. But is that really all you can get for online shopping? I wanted to see if there are really good deals: everybody says you have to go online for the best deals. But I can get better deals than that at the mall! (Mindy Jean)

Mindy Jean, as a Traditionalist, expresses that new ICTs and applications simply cannot match the "old way" of doing things. As she says, she can find better deals at the store. Traditionalists find their new devices repeatedly fall short of their expectations. Compared to their older ICT forms, newer ICTs simply cannot compete:

> My computer I use once in a while. I like to play games on it. That's about it. It's ok. It's not as nice as the TV. My kids bought the computer for me. Otherwise I wouldn't have one. Why would I want a computer when I already have a TV? (June)

As June states, why would she want to use a computer when she has a perfectly good TV? The television helps her fill her days, provides companionship, and helps her experience nostalgia. Traditionalists simply love the ICTs of their youth so much that there is no room left in their lives for more modern ICT forms.

Traditionalists: The Keepers of Technological Tradition

Traditionalists deeply love the ICTs that were available in their youth and young adulthood. In many ways, they carry on the ICT use patterns and traditions they learned in their childhood into elder age. One can think of Traditionalists as keepers of the technological traditions of their youth. Key points about Traditionalists include:

- Traditionalists have very fond memories of their ICT use as children and young adults and still have strong preferences for these ICTs and media.
- Friends and family members gift their Traditionalist loved ones modern ICTs but these remain mostly unused by the Traditionalist.
- They place the older forms of ICTs they love prominently in their homes, hiding more advanced forms.
- While Traditionalists love the ICTs of their youth, they are largely indifferent to newer forms.
- To appeal to a Traditionalist, it is important to evoke generation-specific feelings of nostalgia.

While Traditionalists embrace or reject ICTs based on their relative age, the Guardian user type views all ICTs with suspicion, as we will see in Chapter 6.

Chapter 6

Guardians: The Technological Resistance Fighters

> I feel that there's a need, there's definite need for all this modern technology. There's a need for it but I think it's just like many, many things it's overdone. I think it's absolutely mind-boggling ridiculous that cars now have TVs in them [...] I believe a happy life is a life of moderation. Yes, use a computer, use a cell phone. The cell phone is wonderful because you can be in the grocery store and realize that "oh gosh do I need that?" And you could call home and say, "can you look in the cupboard and see if I need such and such or" or you can call somebody and say, "I'm running late. I got caught in traffic." But you go to the mall and you see people walking around and they're just talking on the cell phones. Talk, talk, talk on the cell phone. I thought you went to the mall to go shopping. So, I think that people go overboard on all that stuff. (Margaret)

Gluttony. Laziness. Waste. Excess. Morality. Boundaries. Control.

Guardians are deeply concerned about the impact of technology on our societies. ICTs are not seen as negative, but rather seen as enabling individuals to wallow in negative traits we all possess: laziness, gluttony, waste, and self-isolation.

Guardians believe that individuals need to set strict boundaries, carefully controlling their ICT use. While deeply concerned about issues such as information security, privacy, and media bias, they use a mixture of ICTs in their everyday lives. Guardians tend to not differentiate between kinds of ICTs in their potential to cause moral decay nor do they make distinctions based on the relative age(s) of the ICT(s). Abuse of the telephone is no different than abuse of the computer in Guardians' minds: both can be used in ways that impact our safety and privacy, and cause harm to people and society.

Guardians are introduced to ICTs by family members and through work. They heavily regulate their own technological use (setting time limits, for instance) and tend to hide all ICTs in their homes to prevent "mindless" or unintentional use. Guardians often have been shaped by extremely traumatic experiences with technology in their early to mid-adulthood.

Formative Experiences

Guardians can have very nostalgic views of ICTs from their youth, much like Traditionalists. Unlike Traditionalists, however, these views are seen in strict contrast with the current time period. Guardians believe that ICT use in the past was heavily self-regulated by individuals. The use of ICTs in the present day regularly isolates, separates, and enables negative traits, such as laziness. Margaret, whom we heard from at the beginning of the chapter, shares how media consumption and ICT use were family focused and social when she was a child:

> Back then, we watched TV as a family. We sat down after supper with a bowl of popcorn and you watched TV. I can remember long before we had a TV, we used to go to the movies. And we had a little movie theater that was within walking distance of our house and my mother loved movies. And so, many a night after dinner, we would go to the movies. We'd go to the movie as a family. Then on Saturdays, as long as we had our chores done, as long as our rooms were picked up and we had helped mom do whatever we were supposed do, Saturday afternoons was the movies. And it was a social thing. Everybody was at the movies on Saturday afternoon. It was where you met up with all the kids that you went to school with, and it was a social thing. (Margaret)

Margaret loved going to the movies with her family and became a lifelong movie lover, like her mother. She often spoke wistfully about her childhood experience with the television and the movie theater. In this and other recollections, however, she makes it clear that her nostalgic memories are in strict contrast to the ICT use patterns of today:

> I think the TV now, at least my experience with TV; even back when I still watched it once in a while, it was more like you're doing your own thing. It wasn't a social thing anymore. It's isolating. The television has sucked people into thinking they need, I need this, I need that. It's become more and more "let's get our Christmas shopping done in September!" Let's almost forget about Thanksgiving, that Thanksgiving even exists anymore. Other than it's Black Friday, the day after Thanksgiving. Let's just run out and buy a whole bunch of stuff that nobody needs. "Oh, I want that." "Where can I get one of those?" What for? TV glamorizes everything: violence, waste, excess. (Margaret)

"Back then" (during Margaret's childhood) television and the movies were time-defined events (such as watching a single program after dinner), family-focused (with the entire family watching one program or movie together), and social (with interaction among family and friends). "Now" television (and other

ICT use) is often time-undefined (with some people turning on the television as soon as they get home or leaving it on all day), individual (with individual members of a family consuming different media), and isolating (everyone using their own devices, people being manipulated to gluttonous consumption, and programs glamorizing violence). Guardians often see a decay in the moral compass of society and they believe that ICTs, if their use is not strictly controlled by the individual, facilitate such decay. Guardians have experienced a highly traumatic, transformative experience with ICTs that occurred before or during their mid-life. These experiences, such as job loss or divorce, are coupled with the use of ICTs in a way that harmed the Guardian. For instance, George felt his lack of IT skills was one of the reasons why he was encouraged to leave the workforce, retiring earlier than he originally had planned. Oftentimes, these traumatic experiences have been tied to the introduction of new ICTs into the Guardians life, be it in their work or families.

Introduction to ICTs

Some Guardians were first introduced to ICTs in work, while others were primarily introduced by family members. However, unlike other types, the introduction of new ICTs into a Guardian's life often was coupled with many negative consequences or correlated with traumatic concurrent events. Margaret shares how the introduction of the computer in her workplace impacted her:

> When I left work I had a computer on my desk. Everybody had a computer on their desk. When I first went [back] to work in 1980 the computers that we had were the computers where you only had incoming information. I don't know how to explain it any other way. But we had no input. It was information that came to us. Most of us in the office when we got "the new computers" were panic stricken. (Margaret)

Margaret was "panicked" when new PCs were introduced to her office. She feared using the computer: that she could break the technology or that she would not know how to complete a task. She and a friend took computer classes together to help her overcome her fears. Other Guardians were not introduced to computer technology in the workplace. Natalie and George were white-collared professionals (Natalie was co-owner of a biological testing company; George was a Vice President in pharmaceuticals). Both avoided computer use due to their positions in their respective companies. (Dan, a Practicalist, discussed in Chapter 3, was in a similar situation.)

Computers, early in their introduction, were seen as secretarial work by many white-collared professionals (Mandel, 1967), and both Natalie and George shared this sentiment. As Natalie expressed, when it came to computers: "I had my secretary do it. We had a secretary – that was what she was there for. I didn't need to know what she did – we could hire it." George shared if he needed something from the computer, "I just had my secretary do it." Much of his work involved using reports that were often prepared by others, including computer technicians, and he shared his work philosophy on computer use had been: "I wasn't running down saying how does this computer work or why do you do this – it's like I didn't care – that wasn't my job."

In mid-life, Natalie's marriage slowly began to sour and she suspected her husband was committing fraud in their shared business. As co-owner, Natalie was extremely concerned that she could be held liable for her husband's deceit. She became determined to learn to use the computer – to investigate her husband's business dealings – secretively. With the help of her cousin, she installed a key stroke tracker on her husband's computer. The key stroke tracker not only confirmed her suspicions that her husband was defrauding the business but also indicated that he was romantically and sexually involved with other women:

> I didn't use a computer until 1999, when I wanted to find out if my husband was cheating the business and if he was cheating on me. So, my cousin would talk to me over the phone: "Now you do that, now you do that." I was scared to death of breaking the thing. It was a step-by-step-by-step-by-step thing. Because I couldn't just jump in and start clicking on stuff! Installing the tracking software gave me some confidence, I got some experience. I was forced to do things. I went into hackers' chat rooms looking for a key stroke tracking program. I called one kid – he was a college student. He had a key stroke tracking program out there. I downloaded it, but I couldn't use it, because you had to know how to use a computer. So, I called him after I asked him for his phone number. I said, "I couldn't install your program I'm not – I'm not computer literate – I'm a housewife. I'm not a computer operator" so he put together a program that was simple. And it did its own thing. Just for someone who didn't know how to run a computer. I used that and it worked. That's how I found out that my husband had a girlfriend – he emailed her so I wouldn't know. (Natalie)

Natalie's first set of experiences using a computer was extremely traumatic; eventually what she found from her installation of the key stroke tracker led to her divorce and loss of her business. She recognized the computer was not the source of the trauma (in fact, Natalie felt quite empowered and confident after she learned a bit of computer knowledge). Rather, this technological learning experience became closely associated with the traumatic experience of losing her life partner and work.

Traumatic past experiences with ICTs have led Guardians to be cautious about adopting new technologies for fear of potential consequences. Many of their new ICTs tend to be gifts from family and friends. Some of these gifts are welcomed, but many are not. Margaret shared that when she retired from her position as an administrative assistant in a financial firm, her work colleagues bought her a cell phone, although they had originally proposed buying her a computer:

> Actually, the only reason why I had a cell phone when I retired is that people at work asked my kids "What could we get her?" Could they get me a computer? My daughter said "no, no, no. No don't get her computer. She won't use it. Get her cell phone. If you have any money left over, get her a nice phone, and just pay part of the plan." (Margaret)

Margaret, as she shared, would not have used a computer if her workplace had bought her one when she retired. She was leaving work, in part, because of her traumatic experiences with computer technology. Instead, she was gifted a simple cell phone; which she felt was a more useful and appropriate gift as she had fewer negative experiences with the cell phone. (Margaret purchased her own computer under the guidance of a neighbor several years after retirement.) Since Guardians concerns focus on ICTs being used in inappropriate ways, they tend not to have issues with receiving ICT gifts that match their values. For instance, Jackie was concerned about the influence of corporations on society, including issues such as monopolization and price gouging. Her now-deceased partner had purchased her an Apple laptop, after several bad experiences with Microsoft in the workplace:

> My husband knew I always wanted an Apple computer because I heard if you're into photography or any of the arts that's the computer to use. Now I had never used it or knew anything about it. I just had too many people tell me that was user-friendly and I hated Microsoft. Quite frankly, I hated it. It used to do whatever it felt like doing not what I felt like doing. But I love my Apple! It was a wonderful gift! (Jackie)

Guardians can be cautiously accepting of technology, particularly if they are convinced that it is secure. But they are very careful about ensuring they use these devices appropriately in ways that are non-damaging to society.

The only ICT which drew universal heavy criticism (and was seen as overall a waste of time) was video gaming. Most technologies, however, occupied a middle ground: if an individual could prevent becoming engrossed in using an ICT, prevent themselves from losing basic manners, and maintain a balanced life, the ICT was seen as potentially good for the individual. Guardians tend to view their own use as appropriate and examples of "good" (as opposed to "bad") technology use, because they carefully self-regulate and set limits.

ICT Use

Guardians carefully structure their ICT use around face-to-face (non-virtual) time with friends and family. They view such face-to-face interaction as being critically important; believing that using ICTs to communicate devalues relationships. In their minds, all people should set and keep goals for incorporating as much non-virtual communication into their lives as possible:

> Technology is often easier than spending time with someone in person. You don't have to put up with bad characteristics and bad habits. I work at having face-to-face time. I make sure I see all my friends face to face, I don't just live in the virtual world. (Natalie)

Guardians, like Natalie, often view choosing digital or virtual communication as shirking away from the work of maintaining a face-to-face relationship. Keeping "virtual" relationships is, to a Guardian, the "lazy" way of maintaining friendships. Jackie shared this preference for physical face-to-face relationships as well:

> I prefer to meet people face to face. I preferred to meet you [the researcher] face to face. I can judge people better face-to-face than over the phone or online. You're a real person. I'd much prefer to spend a half an hour with a friend and see them than spend an hour with them on the phone. (Jackie)

Guardians place a high value on physical presence in their relationships, which stands in contrast to the other types we have discussed. Many Enthusiasts, such as Alice and Fred, had "virtual" friends they had met on messaging boards. Guardians would likely dismiss these virtual relationships as being "less real" or of lesser value than those that were based on physical contact. Socializers, such as Gwen, viewed their ICT use as strengthening and deepening their relationships. Guardians would suggest technology use instead separated and weakened their relationships. Traditionalists, with their heavy ICT consumption, such as Mindy Jean, showed a strong preference for using a phone over email, believing email lacked important social cues. Guardians would suggest that phone and email both lacked social cues and that spending more than a few minutes on the phone with a person was gluttonous.

Guardians limit how many life contexts they use technologies in. They own a device or application for a specific purpose and they carefully regulate their use to ensure it remains on target. Most Guardians could easily identify how they used various ICTs, be they cell phones, televisions, or computers. Natalie shares how her cell phone is used primarily for emergencies, while Jackie shared how she used her computer to check news and communicate via email:

> I think the cell phone is a great thing. Especially in the case of emergency when you have to get a hold of somebody. It leaves you free to perform some task without having to worry about missing a phone call. It gives you little bit of freedom. (Natalie)

> What is important to me is to check the news every single day. I don't have a TV and I don't ever intend to have another TV.

> Therefore, I need to know what's going on. I need to know what's going on in this world as far as news. That is important to me. I feel like I'm lost when I don't. So, I use the computer for my news. News and I do my emails. I check those during the day. That's important to me because that's one of my primary ways of communication with the outside world from home, as I don't like to talk on the phone. I also use the computer to work with my pictures. It's all leisure stuff. (Jackie)

For Guardians, their use is specific, targeted, and regulated. At first glance, this seems eerily similar to Practicalists, who also target their use toward function. However, there are some very important differences. Guardians focus on regulating their use of ICTs to prevent bad habits and traits from being exposed, such as being "sucked in" to watching too much television or spending all day on the computer. Their primary concern with keeping their use targeted to specific purposes and tasks is to prevent absorption and isolation, potentially negative consequences of ICT use. Practicalists target their use toward function and completing a task, viewing the ICT as a tool to get a job done. An example of this difference is that Guardians see the television as a device that "sucks" away valuable family time, while Practicalists view the television as a potential tool of leisure to be used during family time.

Guardians are concerned about watching too much television, spending too much time on the phone, and texting too often. ICT use is viewed as coming at the expense of other, more worthy and important activities:

> Cell phones can be as annoying as all hell because people don't use them as they should. People get those things stuck in their ear [wireless ear pieces] and they're talking with no notice of where they are. People have forgotten about courtesy. I was with a friend yesterday and she was texting. And, of course, the phone kept ringing, and she wouldn't turn it off because she's got family members that might be trying to get to her. So, you couldn't even carry on a conversation. This thing kept making this noise – it got to be an annoyance. In church yesterday, two phones went off. There is a time and place for these things, and it isn't church. (Natalie)

For Guardians, the people directly in your presence require more thought than people who might contact you from afar. As Natalie suggests, Guardians believe that there are appropriate uses of cell phones – but church (as an example) – is not one of them. ICTs were particularly seen to pose a potentially damaging risk to young people. Margaret shares that while ICTs are needed, they are often overused:

> I feel that in life to today there is a definite need for all this modern technology. There's a need for but I think it's just like many,

> many things it's overdone. "It's a beautiful day outside." I used to say that to my grandson when he used to come over. When he was 10, 11, 12 years old, I'd say "it's a beautiful day go on ride your bike. Go ride your bike go out and play. You're not going to sit in here and play that video game all day. If it's raining you can go play your videogame." But there are other things to do. Go out and look at the trees. I think it's absolutely mind-boggling ridiculous that cars now have TVs in them. Look out the window. Enjoy the view, see what you're seeing in the car. It's removing them from a part of life that I think is important. I think it's important to sit down and have a conversation with mom and dad. And talk to grandma and grandpa if you're lucky enough to have grandma and grandpa. (Margaret)

Overuse of technology is seen as removing the individual from the real and important world, the world that, in Guardians' minds, teaches life lessons and forges relationships. Guardians tend to be particularly concerned about children's overexposure to the virtual world, as children have not learned their own self-regulation, as Margaret speaks about. Guardians carefully self-regulate their own use, including how much they use specific ICTs (often setting a time limit):

> I'm not saying that you have to have your nose in a book all the time, but you don't have to have your nose in front of the computer all the time either. I believe a happy life is a life of moderation. Yes, use a computer, use a cell phone. The cell phones wonderful because you can be in the grocery store and realize that "oh gosh do I need that?" And you could call home and say, "can you look in the cupboard and see if I need such and such or" or you can call somebody and say, "I'm running late. I got caught in traffic." But you go to the mall and you see people walking around and they're just talking on the cell phones. Talk, talk, and talk on the cell phone. I thought you went to the mall to go shopping. So, I think that people go overboard on all that stuff. (Margaret)

Guardians believe that technologies are a part of everyday life; but they are *just a part*. As Margaret suggests, technologies have very well-defined uses in life and are convenient (hence why Guardians use them), but they are just "a part" of what should be a diverse life of moderation. If a Guardian feels an ICT is too burdensome to use, leading to an overabundance of waste or decay of morals, they will often choose to stop using that ICT form altogether. Jackie shares how she stopped using her television after her partner's death:

> I used to have a TV. What I found was that we had two TVs. My husband had his and I had mine because we had different things

> that we wanted to watch. Again, I don't like wasting my time so why would I watch something that he wanted to see but I didn't? So, we had two TVs, and I would watch my TV upstairs when I went to bed, it was in the bedroom [...] After a while I found there was nothing I was interested in. All the shows are low quality and the news lies about everything. I won't have a liar in my home [...] Quite frankly I think society as a whole spends way too much time on TV. They should really turn the TV off and do other things. So, when my husband passed I just didn't see any need for anymore. And I had stopped watching a lot of it. I'd go for walks with my dog. I'd read. I'd visit with friends. (Jackie)

Jackie had carefully regulated her own television use by shutting it off and "doing other things." She merely, at best, tolerated using the device, finding no joy in its use. Keeping a television after her partner died made little sense: she no longer watched it and she felt that the content (particularly, the news) was filled with bias. She did not want to welcome such media into her home and when she downsized to a small apartment she did not bring the television.

George, a Guardian, was married to Mindy Jean, a Traditionalist we met in the previous chapter. When he was working full time, he was effectively out of the house five to six days a week for 10 hours or more; which meant that there was little conflict over ICT use and non-use in their marriage. When George retired, Mindy Jean's love of her traditional forms of media often upset him, as he suddenly was exposed to her nearly constant habits of having the television or radio on. As a Guardian, he preferred to limit his time-consuming media or using technology. Being married to a Traditionalist, this was not possible. Mindy Jean quickly became tired of him being critical of her technology use, so she encouraged George to find part-time employment. George quickly took a position with a big box retailer. This allowed Mindy Jean to consume and use her traditional media and devices while George was out of the house, but also allowed George to restrict his exposure to unwanted ICT usage.

For Guardians, their use of ICTs is carefully regulated and guarded. Technologies are not viewed in terms of what they can facilitate (such as tasks or relationships) but on how they can damage or destroy. Guardians place ICTs in their homes in arrangements that discourage their use, reflecting their desire to carefully self-regulate.

ICT Display

Guardians' homes are notable, not because of the ICTs that are visible, but because of the ones that are not. Their homes are remarkably devoid of ICTs in the main living areas. Restricting easy access to technology is a way that Guardians prevent overuse and regulate and control their own technology habits.

Margaret lived in a ranch home with a finished basement. She spent most of her indoor time on the main floor, which was ICT-free, with the exception of her cell phone and a television in the smallest guest room. Visitors could easily spend a week or more without encountering a television, radio, or computer. Margaret prided herself on her ICT-free living room (Figure 7), which she felt facilitated playing piano music, reading and, most importantly, conversation.

Entering Margaret's home, one immediately sees her technology-free living room. Contrast this with an Enthusiast's home, such as Alice's: when one entered, one immediately saw Alice's television, stereo, and laptop. For Guardians, such as Margaret, restricting the positions of ICTs prevents their overuse and, therefore, helps Guardians self-regulate:

> I'm not a big TV fan. I think it's become such a way of life; it's the first thing people do when they walk in the door is either turn on the computer or turn on the television. And it's not like, "how was your day in school today?" "Gee it really smells good in here it looks like you made something nice for supper." Boom, the TV's turned on when they come in the house. Sometimes I think with some people it's just like background noise. But the TVs got to be on or life's not going to be complete. I had an elderly gentleman come here for Thanksgiving. And he came in and he said "oh, this is wonderful, a real living room, no TV. And I said "oh, there's a TV back there in the den." But that is where the TV belongs. (Margaret)

Figure 7. Margaret's ICT-free Living Room.

Margaret is quite proud of her ICT-free living room, and how having such a space prevents her from automatically turning on the television, while promoting the face-to-face relationship building that Guardians find so important. While Traditionalists place the ICTs they love in the center of their rooms and Practicalists place ICTs in rooms based on their function; Guardians hide all ICT forms and place them in difficult to reach locations to restrict their use. Margaret kept her computer, a flat screen television, and a stereo in her finished basement. While the finished basement was quite nice, it was removed from her main living floor and, as a result, she only went to the basement when she wanted to use technologies. Such placement prevented her from "mindlessly" turning the television or radio on for "background noise."

Guardians would often say "electronics shouldn't be in bedrooms" (Jackie) or "a living room shouldn't have a TV in it" (Margaret). In their considerations of where to place an ICT, the focus is not centered on where ICTs *belong* in the home, but rather where they *do not belong*. Guardians work extremely hard to keep their ICT-free spaces devoid of technology, even if it means that their partner or other individuals find their own ICT use restricted. Margaret, for instance, had a boyfriend who enjoyed watching "mindless TV" (Margaret). She placed a small cathode ray tube (CRT) television in the smallest spare bedroom on the main floor for her boyfriend to watch:

> My boyfriend has his own TV [on the main floor] that he will watch all his news programs on. He'd be watching mindless TV, because he's the type that when he walks in the door, he turns the TV on. If he wants to watch TV he has to watch it on his own TV. I don't want to see it. (Margaret)

Margaret placed her boyfriend's television in a very awkward spot in a cramped bedroom (Figure 8), restricting his use. The only way to view the television was sitting on the end of the single guest bed, which provided no back support, and likely became uncomfortable quite quickly.

Margaret, who strictly limited her own television viewing to a single DVD movie per week, had a large flat screen LCD television in her finished basement. Her boyfriend, who did not restrict his use, had a much smaller CRT television placed in an awkward corner in the tiniest room in the house. Such placement of these two televisions juxtaposes the differences these individuals have in their ability to self-regulate. Margaret was able to self-regulate her own use of the television, so she rewarded herself with a pleasant viewing experience: a large television in a comfortable room. Her boyfriend, incapable of such strict regulation, was offered a small television and an uncomfortable viewing area. If her boyfriend could not regulate his own use, Margaret would regulate his use for him.

Guardians, however, do not always have the luxury of hiding their ICTs. Margaret had the space to place her technologies on a separate floor and in a spare bedroom. Jackie did not. She had recently been widowed by her partner of

Figure 8. Margaret's Boyfriend's Television.

over 10 years. When she was young, she had moved her family from their north-eastern US city to Canada to escape a financially abusive ex-husband. She only returned to the US city to be close to family and friends when he passed away, several decades later. After she returned she met her partner, whom she called her husband (although they chose not to get legally married). Jackie did not have a private retirement account, as she had predominantly worked in retail and low-level secretarial positions. As a result of her having spent so many years outside of the United States, her work history did not qualify her for Social Security benefits (the US pension system for those age 65 and older).[1] This left Jackie in her early seventies without any retirement income.

---

[1]Since Jackie had not been married for 10 years or more, she also did not qualify for Social Security marriage benefits. Marriage benefits allow a person to either choose to collect their own Social Security pension, or an amount equal to half of their spouses (or former spouses) Social Security pension; whichever is greater. However, to qualify, one must have been married to their spouse for a period of 10 years or greater, even if they have since divorced. Since Jackie had worked out of the country for several decades and had never been legally married the minimum ten years, she did not qualify for any Social Security benefits, either on her own or under the marriage benefits clause (Social Security Act, 2018; United States Social Security Administration, 2018).

Jackie was financially at risk. She had inherited her partner's estate, which she sold to cancel debt and raise money. She made her living by working odd jobs, often in retail, which were difficult to find because "no one wants to hire an old person" (Jackie). She had realized that she would no longer be able to afford to rent her small one-bedroom apartment and decided to buy a pop-up trailer to live in. She planned to head south, where she could live in a relatively inexpensive camp ground for US$50 a week. When I interviewed her, she was selling her last few items (furniture and home goods) to fund her journey. As a result, her living room had been overrun by boxes of items and furniture she was attempting to sell. The only rooms left in her apartment free of these items were the bathroom, a small kitchen, and bedroom. The few ICTs Jackie owned, mainly a digital camera, laptop computer, and a telephone, had to be kept in her bedroom. The placement of these ICTs directly in her living space caused her quite a bit of stress and worry:

> This [bedroom] is my living area. So, I don't really like my computer in the area that I'm living in – I don't like it where I'm sleeping because electronics just interferes with everything. I've read many places it's not good to sleep in or near these types of things. I read that you sleep better if you don't have it near you. I just read that you shouldn't have your cell phone near your bed when you're sleeping. You shouldn't even be holding your cell phone for that long of the time against your head, because people are coming down with brain cancer because they talk on the cell phone. I'm conscious of that kind of stuff. I'm conscious of health. I know all about it and I do things to avoid the problem. The computer's convenient because it's right here so I use it a lot. While we were talking I heard a click and then I knew I had an email. So sometimes I hear that [click] and I'll check to see what it is. I'm more likely to stop what I'm doing and check my email because it's right here and I can hear when I get one. (Jackie)

For Jackie, the placement of her computer in her bedroom meant she checked her email more often and was more likely to interrupt other activities to do so. As a Guardian, this often upset her. She felt that her future situation in the trailer would likely be better, as she would no longer have internet, and would no longer hear notifications from her email. She would instead rely on the nearby public library for internet service, which would allow her to isolate her use of her laptop to only times when she visited.

Natalie, after her divorce from her husband due to his infidelity, became a hoarder. When her former husband and son moved out of their shared home and ended all contact, she responded by slowly filling up their rooms with items. After filling up their rooms, she proceeded to fill the upstairs hallway, the stairs, and later her dining and living room. These rooms were unnavigable, lacking even pathways, as they were piled with boxes and furniture to the ceilings. She moved her bed into her family room which adjoined the kitchen, placing her computer on her former kitchen table (she stood to eat), and her television in a corner of the family room. Natalie stated that she had "no need for a table because I have no family."

Natalie's television was placed in the family room across from her bed. She commented that if she could, she would find some other place for her computer and television, such that she didn't need to be "surrounded" by them, but that she had no other space in her home.

George felt he had little control over how the ICTs were placed in his home. He relayed, "I'm married. My wife decides where everything goes in this house, not me." So while Mindy Jean, his wife, proudly displayed her traditional forms of ICTs all over the house, George felt he had little choice in their placements. However, George deeply disliked her watching soap operas, so when he was home, Mindy Jean retired to the upstairs bedroom to watch them, a compromise they had reached.

Guardians resist using ICTs in ways that they view as negatively impacting society, their relationships, and themselves. This includes resisting societal expectations to have technologies scattered throughout their home and resisting what they believe is societal pressure to constantly be using ICTs.

ICT Meaning

Guardians often can point to one or several extremely traumatic experiences with ICTs that shaped their perspectives on technology. These are life-changing events in which ICTs are seen as playing a significant role, such as technologies impacting job loss or divorce. Margaret experienced the slow reduction of her career due to technological intervention and Natalie the dissolution of her marriage and abandonment of her family. These traumatic experiences tend to occur before or during mid-adulthood.

For Margaret, the workplace introduction of newer forms of ICTs, most notably the television and the computer, shifted her job at a major financial firm from an enjoyable position to one of drudgery. Margaret went from being an "assistant sale representative" where she had power to act on client requests, to "just being a secretary" who had little social interaction and little autonomy. While her title at the financial firm stayed the same, every technological introduction in her workplace represented a decrease in her interaction with others and a slashing of the most enjoyable parts of her position:

> I was the sales assistant to the manager of the branch and part of my job was to talk to people until [the manager] was ready to see them because he might have somebody else in his office. So, I developed a very good rapport with people and sometimes they'd say, "Well we don't really need to see him anyway, this is what we want to do" and it would be taken care of by me. But

> then there became a time after everything became computerized that I was, more or less, told by somebody higher up than my boss: "You're spending way too much time with people, just get the order, get whatever it is that they want, and then get them out of here."

Margaret's work once involved a high amount of interaction with clients. As the person at the front desk, she was responsible for speaking with them while they waited to see her supervisor and could even help clients to make their trades. However, after the introduction of the computer, a higher supervisor told her she was spending too much of her time interacting with clients; instead, she should take their order and have them leave. As a Guardian, who deeply valued face-to-face interaction, this ran against many of her core values. The organization made other significant technological changes that impacted the office and her position:

> We used to have an old ticker tape up in front of our office. On a daily basis we might have 30 or 40 people that would come in every single day to watch the ticker tape. Some of them would just sit up there, have their coffee, check the ticker tape, maybe put in an order here and there, but it was like a gathering place. Well, then they [the management] took the ticker tape out because it wasn't generating any business according to the higher ups. They didn't understand that it was a community gathering place and it did bring in business. So, then they put in a TV after they took out the ticker and some people came in to watch the ticker tape go on the bottom of the TV, and eventually the higher ups took that out too.

Margaret's workplace first took out the relatively less-invasive information technology of the ticker tape, replacing it with a television. Upon removal of the television, her office no longer was a gathering place, and no longer encouraged non-virtual interaction. Without clients stopping by the office to visit, Margaret lost the face-to-face client interaction that made her job worthwhile. Her organization also decided that the vast majority of communication would be done only through email and not on the phone or in meetings:

> Then after a while I was told "you can't be giving out quotes over the phone unless they have an account here, or unless they've done business within such a period of time." It went from being kind of like a semi-family friendly place to work to being just a place to work where there was more stress [...] Once we got email, it was just [an unwritten policy of] "send me an email" rather than calling. Then people just became just an email rather than a person.

Unlike Socializers, who view modern forms of ICTs as drawing them closer to others, for Guardians, the introduction of ICTs into relationships degrades them. Margaret shared that she felt that the atmosphere of her office had changed and that the increasing introductions of ICTs set employees against technology:

> So, we lost that personal contact and that's the thing I don't like about it. Now I know that's not all the computer's fault, but it was just a whole mindset of instead of that one-on-one human contact being the priority – the computer was the priority. It's cold, it's sterile, there's no feeling of camaraderie. It became man against the machine. I'm not against progress you know, we need to have progress, but I think it's just gone too far. (Margaret)

Traumatic experiences that shape Guardians are not limited to the work environment but can also be of a personal nature. Natalie shared how she felt technology played a major role in the loss of her family, creating distance between her and her husband and son:

> There were two men [my husband and son] so two TVs in our house. So, I'd be upstairs sewing or listening to the radio or music on a tape that I had. They wouldn't let me watch anything I wanted to watch. They'd watch TV for eight hours a night and wouldn't talk to me. I can remember when we got our cabin. The cabin was to get my son out of the city – out of the suburbs. We went up there and we only got one channel on TV. So, my husband and son would drive seven miles to a convenience store and rent DVDs – a video. They wanted a dish on the roof and I put my foot down and said "No, this place is for us, not for TV." We bought that place to be a family and save our marriage, not to sit there mesmerized by some mindless program. I had my sewing machine, and their TV went in the next room. For some strange reason my sewing machine made snow on the TV. So, when they were there I couldn't sew. So, I was glad when they weren't there – all they did when they were there was watch TV and not talk to me anyway. At least when I was alone I could sew. Because what's more important – you sit there and watch TV or you spend time with your family? (Natalie)

For Natalie, the television resulted in her family life being disrupted. In their home, she found that her husband and son watched television programs that she did not enjoy. Often, her husband and son watched different programs simultaneously on different devices, placing distance between them as a family. When Natalie and her husband's marriage began to collapse, they decided to buy a cabin a few hours travel away. The purpose of the cabin was to be able to spend time together as a family, away from the distractions and influences of "city life," in an attempt to heal their marriage. When her husband and son wanted to have a dish installed to watch television, she refused, and they chose instead to rent videos against her wishes. Her family failed to connect, even when they went to the lengths of buying property and taking vacations to do so.

Since the one activity that Natalie greatly enjoyed, sewing, interfered with her husband's and son's use of the television, they eventually told Natalie she could no longer sew when they were present. Essentially, in her mind, her husband and son chose television watching over spending time with her and even prohibited her from partaking in her hobbies.

Natalie's experiences with technology went even deeper, however. As mentioned earlier in the chapter, she first learned to use a computer when she became convinced her husband was defrauding their shared business. However, she had attempted to learn to use a computer previously, and she asked her son for lessons:

> I asked my son to teach me to use the computer, but he said I was too stupid to learn. So, after my husband and son moved out I was free to use the computer, the TV […]. I finally got to use these things! Because they had monopolized everything – that is all they cared about – the TV and the computer – not me! (Natalie)

One can imagine the hurt that Natalie felt, as a mother, to be told by her own son that she was simply "too stupid" to learn to use a technology. For Natalie, technology use represented the traumatic experiences in the breakup of her family. The television separated her from her husband and son; the computer hid her husband's extramarital affair and business impropriety; and her own son berated her over her lack of technological knowledge. Natalie felt that her husband and son loved their ICTs more than her. Eventually, she would become estranged from her son, having not spoken to him in nearly a decade when we met.

Guardians also have concerns about how ICTs are used by organizations and governments, believing many institutions are using technology nefariously:

> I think that there's a lot of abuse on the Internet and it's coming from big companies. That's the worst part. We all know that the population has a certain amount of criminals in it. But you don't expect big corporations to be part of those criminals. And they are more and more. I could go on; we'd be here for 10 hours […] if I went into all of it! [Laughs.] My computer is what allows me to be aware of this manipulation by the big companies, because I won't watch TV. TV is just a bunch of brainwashing. The TV lies, and I abhor lies. Those TV news people will actually lie. They literally look into the camera and lie. (Jackie)

Jackie, and all Guardians, see good in the respective ICTs they use: the internet has a purpose for researching the truth and learning the latest news. However, these devices can also be used for negative purposes: from removing people from present relationships to outright lying to, manipulating, and misleading people (such as the television). For Guardians, ICTs represent a sort of Pandora's Box: they hold both good and evil and, unfortunately, they feel that society prefers to delve into the evil and use ICTs for negative purposes. *It is not the device that is negative, but how it is used.*

Guardians have strong beliefs that much of society lacks strong morals and good boundaries with regard to ICT use. They see many individuals struggling with how to effectively integrate technologies into their lifestyles in balanced ways. Instead, many are overcome by their absorption into the virtual world, leading to a displacement of their "real" (physical and face-to-face) relationships:

> I see these technologies as taking up a lot of time that could be spent actually living rather than watching. Especially things like 3D where you participate in a world that is generated instead of the real world. To me it's a copout. It's also very "one-manship:" it doesn't include other people. It's very isolating. They're replacing human relationships with technology. It's easier to do that than talk to someone. Talking to someone takes work. Using the technology is more convenient. It's always available. You don't have to work at relationships. Just push a button. (Natalie)

Considering the extremely traumatic experiences Guardians have often had with ICTs in their work or personal lives, it is not surprising that they view using ICTs as an easy alternative to maintaining face-to-face relationships. It was easier for Margaret's workplace to treat her as "just a secretary" and have her send emails than to have her (from the organization's point of view) "waste" time establishing relationships. For Natalie, it was easier for her husband and son to watch television and play videogames than to work on repairing their family life.

For Guardians, it is important to set strict boundaries to prevent being absorbed into the virtual world, or to prevent taking a "copout" (Natalie) from the "real" one. Guardians talk at length about the importance of setting these boundaries around their own use, but they also believe all society should set better boundaries:

> There are other things to do besides technology. There is the old adage that there's a time and place for everything. It's just like everything else. It's kind of like when kids are little and you set boundaries on dinner time. If I make something new, you have to try it. If you don't like it, fine, I won't make it anymore. But we got to have something where we have a balanced meal. We will sit down and eat dinner together. You're not going to go in the

> den with your little plate so you can watch some television show. We're going to sit down and have a family dinner, because this is our time to share. This is our time to discuss what you did today. This is our time to get together and socialize. It's not going to be just come in, grab a bite to eat, something, whatever is in the refrigerator, and then go lounging in front of the TV until bedtime. That's not going to happen. I see young families today that use a certain amount of constructive discipline and I see that they do much better with their children than the ones that just let them do whatever they want to do. Control the family. But it's not just technology – it's just life. (Margaret)

Margaret's dinnertime analogy reflects Guardians' views that ICTs are best used in moderation. New technologies should be tried (much like when your mother makes a new dish); however, their use must also be moderated (much like how parents set rules on eating at the table). Teaching children moderation in all parts of life is important to Guardians. Technologies are very much like unhealthy snacks: having a few in moderation is fine, but being gluttonous is unhealthy.

In an increasingly technological society, Guardians view their resistance to inappropriate ICT use as activism. As activists, they avoid purchasing or using ICTs which they believe violate their privacy or encourage corporate greed, believing that their actions can help to create a subculture of positive ICT use. Guardians view wasteful ICT behaviors as similar to other wasteful behaviors in our societies, such as food waste:

> Really, I find it offensive when I see the waste. I grew up in the Depression and the War and I find it offensive that people today don't understand want versus need. I go to people's houses and they are wasting food… and I see people waste technology too. Sometimes I'll see someone with a smartphone and I think "gee I wonder if I should really get one of those" and I go, "no, because you would never use it to the capacity, you don't need it and you don't really want it, and if you think you want it it's only because somebody else is telling you should want it." People don't understand the difference between a want and a need anymore, and all this technology stuff does is cause unrest. The companies make you want it. People are wasteful throwing all their old technology away. All they want is new, new, new. (Margaret)

As a child of the Depression, Margaret often grew up without basic necessities, including food, and has a strong distaste for waste as a result. She sees many modern families who are wasteful in general, not only with food, but also with technology. She moderates her own behavior, choosing to not purchase devices that she knows she cannot fully use, much like she does not prepare or buy food she knows she will not eat. She resists being influenced by commercials and the media, which she believes tries to convince her to consume more technology. Understanding the difference between a "want" versus a "need" is a fundamental skill that she feels society has become inept at, and this includes understanding "want" versus "need" when it comes to purchasing technology.

Guardians believe technological boycotting helps prevent organizations, the government, and corporations from abusing individuals socially or financially. Jackie spoke about how corporations often take advantage of individuals and how she actively resists patronizing these organizations:

> I was angry when I found out my MP3 player wouldn't work because it was too old and they wanted me to buy a new one. It was only a few years old! I was literally angry. When I'm angry I fight back by not buying the products anymore. That's the way I fight back for everything I don't like. I don't go to banks I don't like. I cut the credit cards of banks that have given me a bad deal. I resist. That's what I'm doing, I'm resisting, I'm boycotting. That's what I do when they make me angry. I choose, and I tell everyone I know my experience. So that they also are aware, because that's the only way you're ever going to be effective is that if enough people know it something will happen. If enough people do it something will happen. A company will change. (Jackie)

When Jackie speaks about her experiences with her music player, she couches them in her general purchasing behavior. It is not simply ICTs and their use that are negatively influencing society, but rather it is general trends in organizational practices that leave individuals open to the risks of being given a "bad deal" (Jackie). You will notice that she is actively resisting falling prey to these larger trends, seeing herself as a "fighter" and a "resister."

Whereas Enthusiasts are activists for encouraging ICT use, Guardians are activists for balanced and critical ICT use. Both Margaret and Jackie sought to influence those around them to carefully consider their ICT use and purchasing. Despite Guardians' beliefs that they are resisting inappropriate ways of using technology they often feel isolated and disconnected, misunderstood by the general population:

> I feel like I live in a different era than everyone else. I am content with what I have. I just sometimes think that the younger generation has so much input into their lives that they don't get a chance to just sit down and enjoy life. I'm not a negative person, I just feel so strongly that life is not compromises. Life is so much better if everything is done in moderation. I'm not knocking the wonder of the computers or all the modern technology that we're so lucky to have! I'm just saying that I think it's become an

> obsession with a lot of people. But other people think I'm just strange. (Margaret)

Margaret expresses that she often feels out of place in modern society, a feeling shared by many Guardians. Guardians' feelings toward ICTs are often harshly dismissed by society at large.

Guardians: The Technological Resistance Fighters

Guardians have a deep concern over how ICTs are used in everyday life. While they often use a plethora of technologies, their use of these ICTs is very restricted in terms of both the functions and the amount of time they use them. Guardians actively resist using ICTs in ways they view as negative for society. They view themselves as activists fighting for balance in an increasingly technological environment, as resistance fighters in a digital world. Key points about Guardians include:

- Guardians can often point to one, or a series of, very salient life-changing traumatic experiences with ICTs.
- They believe that technologies should be used in moderation and this use should be carefully controlled. Guardians carefully control and restrict their own use.
- In their homes, Guardians prefer to hide ICTs, placing them outside of the main living areas.
- Guardians do not see ICTs themselves as negative, but rather that technological use brings risks of becoming gluttonous, isolated, and lazy.
- To appeal to a Guardian, technologies must be private, controllable, and secure as Guardians are highly sensitive to privacy and information security risks.

These five user types demonstrate five distinct patterns of domestication in the use, meaning, and display of ICTs. Enthusiasts love ICTs as fun toys, Practicalists see ICTs as tools, and Socializers see ICTs as connectors. Traditionalists love older forms of ICTs but reject newer forms, while Guardians view all ICTs with suspicion. An overview and summary of the ICT User Typology follows in Chapter 7.

Chapter 7

Understanding the ICT User Typology and the User Types

The five user types of the *ICT User Typology*, the Enthusiast, Practicalist, Socializer, Traditionalist, and Guardian, capture the diversity of older adult ICT users from those who are excited to those who are fearful. Each user type represents a distinct domestication (Silverstone & Hirsch, 1992; Silverstone & Hirsch, 1994; Silverstone et al., 1994) pattern in the way that individuals are introduced to, use, display, and come to find meanings in ICTs. People can be categorized into these patterns based on their general philosophy toward technology. Table 1 integrates all five user types, exploring the domestication pattern for each, allowing easy comparison:

Enthusiasts love ICTs, display them proudly in their home, are interested in new innovations, and center their lives around technology. They can often remember "tinkering" with ICTs in their youth, an activity encouraged by a technological mentor who had a deep interest in technology themselves. They often use multiple ICTs at the same time and they push their workplaces to use more technological innovation. Enthusiasts' seek out new ways to use ICTs they own in every facet of their lives and form relationships with others over the topic of technology and through its use. Technologies are displayed predominantly in their homes to allow easy and constant use. Eager to try new innovations, Enthusiasts see ICTs as fun toys.

Practicalists enjoy using ICTs which fulfill a need and help them complete tasks in their everyday lives. They see the ICTs they use as being function-specific and highly tied to a single aspect of their lives, be it for use in their work, family, leisure, or community. These technologies are viewed as for their own personal use and they place a high value on those with proven functionality. They have no interest in exploring how to use a single ICT in every facet of their lives. Practicalists place ICTs in function-specific rooms, often placing computers in home offices and televisions in entertainment rooms. Practicalists see ICTs as tools.

Socializers have large intergenerational networks of friends, family, and community contacts. They use their ICTs to create, grow, and maintain these connections. Technologies which allow them to socialize and build these bonds are highly valued; technologies that are seen as isolating have little value. Socializers, who have busy lives due to high community involvement, prefer

Table 1. The ICT User Typology: Comparison of the Five User Types.

User Type	Views ICTs as	Formative Experiences	Introduction to ICTs	Use	Display	Role in Society in Regards to ICTs
Enthusiasts	Toys Fun Play	Positive Tech mentors Encouraged to tinker	Media Relationships Self-sought	Stretches ICTs across life contexts Simultaneous multiple ICT Use	Every room Focal point	"Technological Evangelists"
Practicalists	Tools Functional	Early work experiences	Work Relationships	Focuses on function Enjoys practical application Single life context use per ICT	ICT-specific rooms Functional placement	"Technological Tool Users"
Socializers	Connectors Bridges Social	Highly intergenerational families and communities	Relationships (especially relationships with younger individuals)	Focuses on connection and community Enjoys socialization, constant contact	Mobile ICTs Moves ICTs with them	"Technological Social Butterflies"

Traditionalists	Tradition Nostalgia	Positive experiences with ICTs in youth	Relationships (gifting of unwanted ICTs)	Enjoys using ICTs of their youth Rejects newer ICT forms	Older forms displayed Newer forms hidden	"Technological Tradition Keepers"
Guardians	Allowing individuals to wallow in negative traits Harmful	Salient traumatic experience(s)	Relationships Work Self	Focuses on control Highly restricted Disciplined	ICTs hidden Restrict use through inconvenience	"Resistance Fighters" in a Digital Society

mobile ICTs that they can take with them or use while completing other tasks. Most importantly, they choose to mimic how their youngest contacts use technology; adapting to and adopting the use patterns of the youngest members of their social networks. Socializers see ICTs as connectors.

Traditionalists love the ICTs of their youth. They tend to be very nostalgic and place a high value on the ICTs that were available when they were young adults and children. Their lives are so full of these more traditional ICT forms that they see little to no value in using more modern ones. They receive advanced technologies as gifts from friends and family members. They will try these devices (often under the urging of their loved ones) but find that these newer technologies cannot compare to their beloved devices. More traditional forms of ICTs are displayed prominently in their homes, while the newer devices they have received as gifts are hidden. Traditionalists love the ICTs of their youth but see little use for newer ICTs in their lives.

Guardians highly value in-person face-to-face (non-virtual) contact and relationships. They view all ICTs (regardless of when an ICT was introduced into their lives) as potentially negative influences: devices which allow people to wallow in gluttony and wastefulness. They believe strongly that technologies themselves are not negative, but rather can be misused with negative consequences for individuals and society. Guardians set strict self-imposed guidelines on how often, how much, and when they use ICTs, resisting what they view as societal pressure to become consumed with and by the virtual world. Guardians place ICTs into areas of their homes which make them less accessible to prevent themselves and others from "mindless" use. They are proud of maintaining ICT-free spaces in which they can enjoy non-virtual relationships and tasks. For Guardians, ICTs are devices which allow us to wallow in our negative human traits, especially if we are not disciplined in using them.

It is important to note that these user types represent the predominant way in which individuals approach ICTs in their lives. Regardless of a person's predominant approach toward technology, almost all people have concerns over its misuse. While Enthusiasts tend to love new ICTs and are willing to try them, most are still concerned about issues such as privacy and corruption. However, the difference between Guardians and Enthusiasts is that Guardians' predominant view is shaped by concerns of misuse, whereas Enthusiasts' predominant view is shaped by a sense of play. Almost all participants shared concerns over the misuse of technologies, regardless of their type. For types other than Guardians, however, these concerns are very much secondary, if not tertiary. For Guardians, these concerns take center stage.

For instance, Alice, an Enthusiast, shared a story of how she found her Facebook profile pictures placed on a family tree website. She was disturbed by finding her pictures posted elsewhere and asked for them to be removed. However, Alice continued to use Facebook, stating that she loved it, but realized that there were risks involved. Other Enthusiasts expressed concerns about issues such as net neutrality and the influence of advertising. However, these concerns were secondary to Enthusiasts' love of ICTs. Given the number of contact hours spent in interviews, it is unsurprising that almost all the individuals in the study

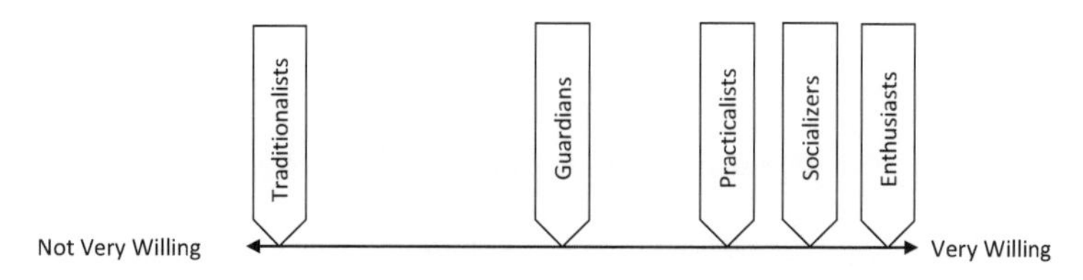

Figure 9. Willingness to Try New ICTs.

shared that they had concerns with technology misuse. Indeed, most of us have such concerns, and if we do not, we probably should.[1]

Comparisons between the User Types: Understanding Similarities and Differences

While each of the user types is distinctly different, there are some similarities between the types. Some types are more willing to try new ICTs or more knowledgeable than others. Several of the types are extremely nostalgic, with fond memories of using ICTs in their youth, while others are not. This next section uses scale gradient bars to illustrate several important features of each type to facilitate easy comparison.

Willingness to Try New ICTs

Enthusiasts tend to be extremely willing to try new ICTs, viewing new technologies much like a child views new playthings, carrying this willingness to experiment with technologies from childhood into adulthood (Figure 9). Anything new, as long as they see it as fun, they actively want to try. Alice and Fred, who were romantic partners, often were trying new ICTs and showing them to each other. Socializers want to try any ICT they observe their social networks using to communicate and pay particular attention to those used by younger individuals. Gwen, as a Socializer, was shown a tablet by her daughter at a sporting event and was immediately interested in trying it. Practicalists are also willing to try new ICTs, as long as the technology has a proven function that they believe

[1]Overall, privacy and information security (and other ethical issues) has been understudied in Gerontechnology (Schulz et al., 2015), despite the ethical concerns involved in many ambient (van Hoof, Kort, Markopolous, & Soede, 2007), assistive (Mortenson, Sixsmith, & Woolrych, 2015), and caregiving technologies (Parviainen & Pirhonen, 2017). As privacy concerns just beginning to be addressed in much of the Gerontechnology literature, the current ethical concerns have focused on the individual (concerns over a person's privacy or agency) and fewer on how a technology impacts a larger social group, such as communication patterns in a care setting. As Coeckelbergh (2018) suggests we, as Gerontechnologists, should begin thinking about such technologies not only in terms of individual impacts, but also societal and group impacts.

will benefit their daily lives. Belinda was constantly learning new ICTs for her job and targeted her exploration of these new technologies for educational purposes.

Guardians, while they are suspicious of all ICTs, are moderately willing to try new ones. They often have concerns about the quality of media content and are concerned about issues such as privacy and information security. If they have adequate reassurance that something is indeed safe, they will try a new technology. They will likely continue using such a technology as long as it is not too intrusive to their non-virtual lives: they can restrict and control its use. Margaret was willing to adopt online shopping after she had reassurance from both her neighbor (who served as her technical support person) and was walked through the process by an employee of the company over the phone. Both reassured her online shopping from this retailer was secure (and explained the security procedures in a way she could understand).

Traditionalists are the least likely to try new ICTs, at least in any sustained way, as they have little interest in newer innovations. They often are gifted these technologies from well-meaning family and friends and may try them under duress or urging. However, their use is not sustained. Mindy Jean's laptop had been used several times in the six months she had owned it, but it was typically stored in a locked desk and remained untouched for weeks at a time unless her children prompted her to get it out for a lesson.

Fun versus Function

Enthusiasts tend to see ICTs as pure fun. They enjoy and expect functionality out of their devices, but technologies are more playthings than work (Figure 10). Fred often spoke about the great fun he had with various technologies from the television to the smartphone. Traditionalists also believe that ICTs are great fun and, therefore, use their chosen devices quite frequently. For Traditionalists, their favorite and most fun ICTs are those of their youth. Newer ICTs simply are not fun to Traditionalists. June found her television fun, but not her computer.

Socializers consider technologies that can be used to build community and/or communicate great fun. Non-social or isolating technologies that Socializers cannot use to communicate or build community are seen as having little value. While Gwen had great fun making up text speak, she had no enjoyment from watching her television, as she believed it was an isolating device. Such beliefs

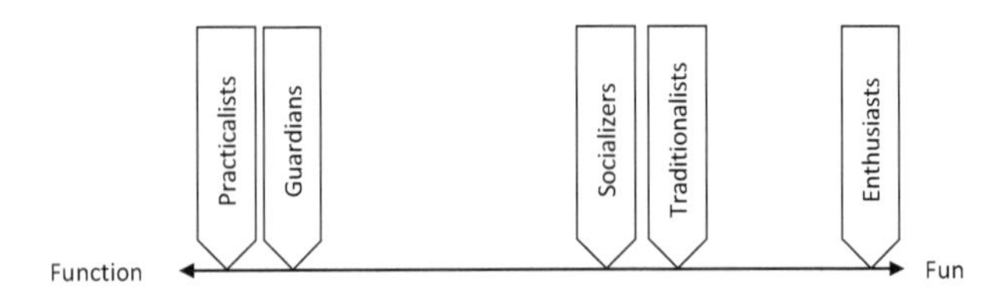

Figure 10. Fun versus Function.

emphasize that Socializers base their judgment of fun primarily on their beliefs about the communication and/or community building potential of a technology.

Practicalists and Guardians view ICTs as much more functional devices than fun ones. Practicalists are concerned primarily with the usefulness of a technology; their interest is in how an ICT's functionality can improve their lives. Many Practicalists, like Boris and Belinda, are intentional about not playing with technology as a toy, but rather using it to complete tasks as a tool. Guardians, who view ICTs as potentially negative influences, tend to not see ICTs as fun, but rather functional items that they use to complete an activity without becoming too absorbed. Margaret carefully regulated her own computer use to a small number of tasks, limiting the time she spent online.

Experiences with Technology (Positive versus Negative)

Enthusiasts have had the most positive experiences with ICTs throughout the life course (Figure 11). They have loved ICTs from childhood, integrated them into every aspect of their lives, and have introduced numerous technologies into their workplaces, families, and communities. Their experiences are overwhelmingly positive and they have positive memories throughout the life course about ICTs.

Practicalists, Socializers, and Traditionalists also tend to have positive interactions with ICTs. While Practicalists and Socializers do not wax poetic about their love for technology as Enthusiasts and Traditionalists tend to, they do tend to focus on positive interactions they have had. Belinda, a Practicalist, spoke about how ICTs enabled her work as an educator. Gwen, a Socializer, would speak about how technology positively enabled a connection with her family. Traditionalists' experiences with the technologies of their youth are overwhelmingly positive. Mindy Jean's discussion of soap operas and how they helped her to bond with her mother is an example of the overwhelming positive memories Traditionalists have of traditional ICT forms. Newer ICTs are met with indifference.

Guardians, however, stand apart in this area, as the only type that can pinpoint extremely traumatic and life-changing experiences with ICTs. Margaret's example of the downgrading of her job and Natalie's family's disintegration, and the part that technology played in both these occurrences, are prime examples of such traumatic experiences.

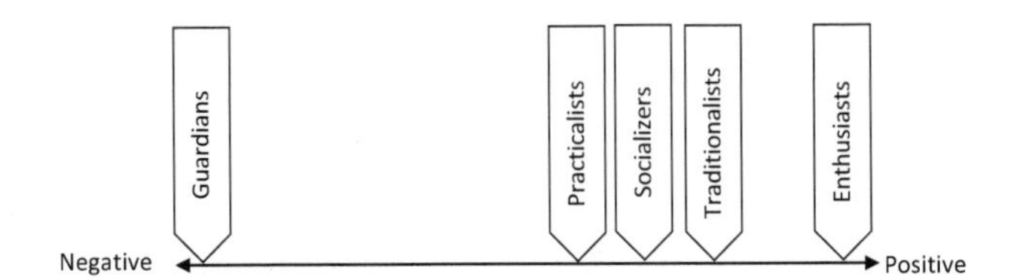

Figure 11. Positive versus Negative Experiences with Technology.

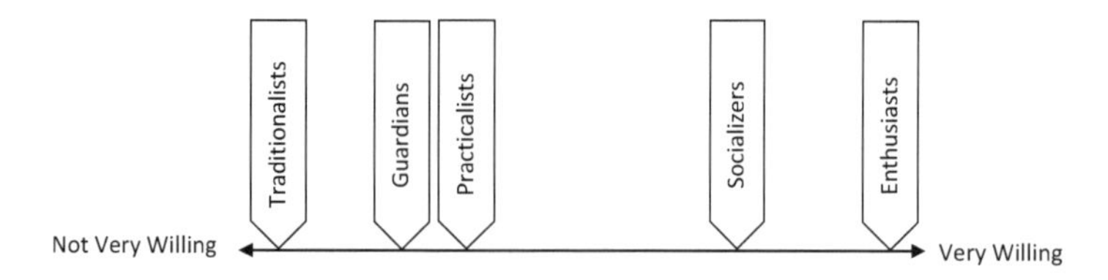

Figure 12. Willingness to Experiment/Play (On Their Own) with ICTs.

Willingness to Experiment with their ICTs

Several user types are much more willing to experiment and discover, on their own, new uses and functions within an ICT that they already own (Figure 12). Enthusiasts love to "play" and experiment with their technologies. Alice, for instance, spoke about how she experimented with her smartphone to find uses in her work as a home health care nurse and then encouraged others in her office to try these new applications.

Socializers are interested primarily in communication and connection and specifically experiment with ways to use ICTs for social, communicative, and community building purposes. Nancy's use of the gaming console in her assisted living center is an example of how Socializers experiment with using ICTS that many would consider non-social to build relationships and community.

Guardians and Practicalists are less likely to experiment with ICTs. Guardians are uninterested in investing a significant amount of time to learn new uses. Jackie shared that while she enjoyed using her laptop for digital photography and news, she wasn't interested in finding other uses – she would much rather go for a walk or visit with friends. In Guardians' minds, the time spent on deep exploration removes them from their everyday non-virtual lives, where they place the most importance. Practicalists are similarly disinterested in experimenting on their own to find new uses, as they believe ICTs are tools, and not playthings. Belinda, for instance, stated that she did not want to "play" with the internet, but rather used it in a targeted and focused way to find educational tools.

Traditionalists, who only try newer ICTs at their families' and friends' urging, have no interest in experimenting with ICTs on their own. June's children had set up a Facebook account for her and would often call her and urge her to "go online." She often refused, telling them to just tell her on the phone. June had no interest in using or exploring how to use Facebook and had never posted her own content.

Technological Fear

Much of the gerontechnological literature has addressed older adults and technological fear/ anxiety (Nimrod, 2018). In the early literature, this was often

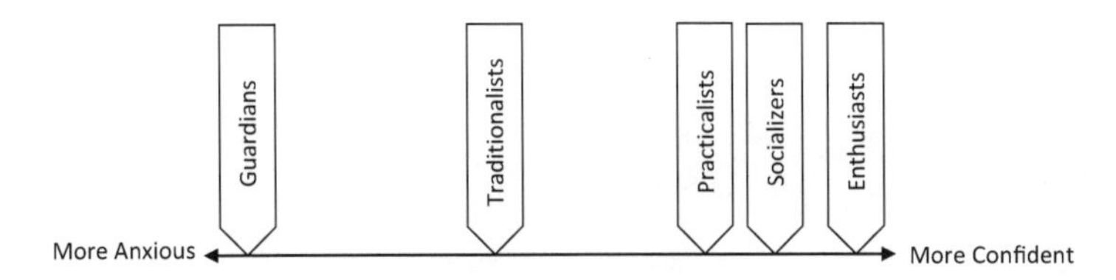

Figure 13. ICT Anxiety.

termed computer anxiety.[2] Numerous studies have shown that computer courses (Dunnett, 1998; Dyck & Smither, 1996) and exposure to ICTs (Czaja et al., 2006; Jay & Willis, 1992) are effective countermeasures.

A different way to conceptualize this anxiety is to term it as confidence or self-efficacy. Prior studies have shown that self-efficacy, or a person's self-belief that they can use a technology, is critical to using an ICT,[3] while feelings of mistrust and worries about privacy and information security decrease use (Golant, 2017; Yusif & Hafeez-Baig, 2016). Enthusiasts, Socializers, and Practicalists express very little computer anxiety and tend to be very confident about their use of ICTs (Figure 13). (A notable exception to this, however, is Practicalists who have not had work exposure to ICTs, such as in Dan's case.)

Traditionalists, who chose not to use ICTs, tend to express neither confidence nor anxiety. They do tend to report lower computer skills due to their choice to not use more modern ICTs in their daily lives. However, it is important to note that Traditionalists do not feel that computer anxiety is the source of why they do not choose to use more modern ICTs. Rather, they are simply disinterested and unmotivated. Mindy Jean shared that she had no desire to use a computer, despite her children's urging.

Guardians show a high amount of technological anxiety. They tend to be overly concerned about internet security and privacy and express fear that they can unwittingly "break" or "harm" things. Much of this concern is overwrought, however. Most Guardians have better technical skills than for which they give themselves credit. Jackie once cleared a computer virus from her machine by calling her local computer store, and following their instructions, even though by her own measurement she was not that technologically competent.

---

[2]See for example Charness, Schumann, and Boritz (1992), Dyck, Gee, and Smither (1998), and Laguna and Babcock (1997), all studies which focus on technological anxiety and older adults.

[3]The literature on self-efficacy and older adults is well developed (Eastman & Iyer, 2004; Haddon, 2000; Hogan, 2005; Lagana, 2008; Lam & Lee, 2006; Lee & Coughlin, 2015; Tsai, Shillair, Cotten, Winstead, & Yost, 2015; Turner, Turner, & Van de Walle, 2007).

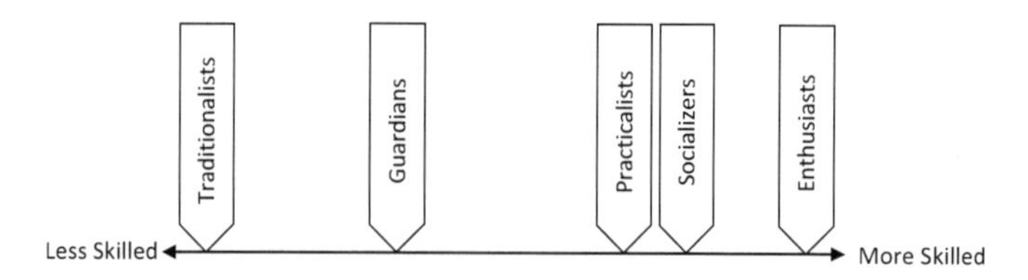

Figure 14. Self-assessed Skill Level.

Self-assessment of Own Skills

Enthusiasts are, by both their own estimates and those of others, the most skilled ICT users (Figure 14). They live and breathe technology and, therefore, their skills match their high levels of use. Enthusiasts are skilled not only for their age bracket, but as could be seen in Chapter 2, skilled compared to the general population. Harry, for instance, was the technical support person for his entire family, many of his friends, and his coworkers – despite being older than the vast majority of them.

Socializers and Practicalists have mid- to upper-level skills, depending on their exposure, although they exhibit more targeted approaches to learning new ICTs than Enthusiasts. For Socializers, their skills are solidly focused on communication devices used within their social circle(s). They quickly explore all the ways others are using these devices to communicate and replicate these patterns. Gwen often spoke about how she learned text speak to communicate, using many of the same acronyms her grandchildren used.

Practicalists tend to be well versed in the ICTs they use frequently. They are not explorers like Enthusiasts and are more likely to rely on someone to train them (often through their work) or illustrate new functions or uses. Since Practicalists are unlikely to explore on their own, their skills outside of their regular tasks can be quite undeveloped. Dan, who worked in a director position in a global non-profit, had little exposure to computer technology in his work. When he returned to the US after several decades living overseas, he found that he no longer had the office support he enjoyed prior to retirement and so he began taking lessons from his wife. This reflects both Practicalists' targeted use (and non-use), but also their resiliency and ability to use new tools they encounter. Practicalists, once they have determined that something is useful, will marshal the resources they need and learn how to use a device.

Guardians and Traditionalists tend to rate their skills as being lower than the other user types. Guardians tend to have a moderate level of skill when it comes to computer use. They tend to slightly underrate their skills, as seen previously, as they believe that they are less competent than they truly are. In particular, Guardians are well versed when it comes to issues such as net neutrality, internet security, online safety, and search engine optimization. Although they may not always know the exact technical terms for these concepts, they understand the basic premises. Their tendency to believe that technological devices can unlock negative human traits makes them sensitized to learning about these issues from the media and from their own experiences. Jackie spoke about how corporations used techniques to ensure that their results appeared first in a search engine (search engine optimization), something she had discovered from her own observation and then extensive research. In many ways, Guardians allow their anxiety and concerns over technology to cloud their own judgment of their actual skill level.

Traditionalists, who are not modern ICT users in their elder years (outside of required work use), tend to have the lowest skills. If they used advanced ICTs in their work they are well versed in using those ICTs; however, if retired, their skills may be dated. June's computer skills had degraded since her retirement from her legal secretary work and although she once had a good basis of knowledge, she did not desire to expand or update it. Mindy Jean, who had been a stay-at-home mother and then wife, had never been exposed to computer technology in the workplace and, therefore, lacked even basic computer skills. Despite her husband's and children's urging (and technological gifts), she did not wish to learn how to use the computer or to use her cell phone for features beyond basic calling.

Nostalgia and User Type

While Traditionalists and Enthusiasts tend to be very nostalgic about technology (Figure 15), this nostalgia is decidedly different. For Traditionalists, they tend to fondly remember using the ICTs of their youth, often reminiscing about important relationships that involved use. Mindy Jean's relationship with her mother, as they watched soap operas throughout their shared lifetimes together, is a prime example of such nostalgia. For Traditionalists, they have fond memories of the media and technology of their youth, and (in part) because of this fondness, they continue to predominately use these technologies throughout the rest of their lives.

Enthusiasts also tend to have fond memories of the technologies of their youth, but these experiences are qualitatively different: Enthusiasts remember playing with the technology, often tearing technologies apart and putting them back together. Most importantly, Enthusiasts remember having a mentor, an important person in their lives who shared their own love of technology. This mentor encouraged them to tinker and play with the technologies they encountered, as Fred's father brought home radios for him to tear apart. This love of

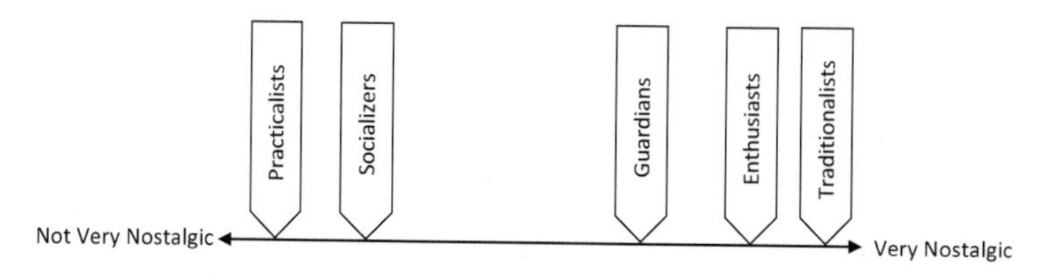

Figure 15. Tendency Toward Technological Nostalgia.

technology and this view of ICTs as playthings and toys continue to be Enthusiasts' predominant view into adulthood.

Guardians tend to have a moderate amount of nostalgia toward the ICTs of their youth. Often, they remember technology use in their childhood as being more moderated and controlled by individuals than in current times, and therefore, more ideal. Margaret shared that as a child she enjoyed watching movies and television as both (at the time) were social activities that nurtured family relationships. Currently, Margaret felt television was used as an isolating activity that degraded relationships. Moderately nostalgic Guardians contrast with Practicalists and Socializers, who are relatively un-nostalgic.

Practicalists, who are focused on the function of ICTs, and Socializers, who are focused on the potential for ICTs to connect them to others, tend to not be overly nostalgic about the ICTs of their youth. In their opinions, modern ICTs provide greater levels of functionality and connectedness (respectively). Such advantages clearly outpace any fond memories they may have.

Overall, the five user types in the ICT User Typology allow us to better understand how older adults approach technology, how they use it, and the meaning technology has in their everyday lives. But how and when do these user types develop? Is there applicability of these user types to younger generations? Chapter 8 addresses these issues, examining these five types in the data collected from the age and generational diverse family members, friends, and coworkers of the Lucky Few participants presented in Chapters 2 through 6.

Chapter 8

User Types and the Life Course: Toward Understanding the Universality of User Types

The ICT User Typology not only predicts ICT use among older adults, but also suggests that these five user types are universal and not generationally bound. While the typology was discovered using a generational (or birth cohort)-specific study of ICT domestication, evidence suggests that these types develop over time as every generation ages. This is illustrated in the secondary data – data from the friends, family, and coworkers collected as part of the investigation of the older adult's ICT use in their relationships.

Data we have already explored from our Lucky Few participants provide substantial clues as to user type development. A review of the retrospective stories told by older adults suggest that a person's ICT user type begins developing in early childhood and is influenced by events through middle age. By the time, an individual reaches elderhood, their user type is set, and is unlikely to be influenced.

But what about technology? Can the introduction of a fantastic new technology change someone's type? The ICT User Typology suggests that the historical introduction of ICT innovations makes little difference. Enthusiasts, who were encouraged to use the ICTs of their youth (which are often quite different from those of their elderhood), remain enthusiastic about technologies through old age. Guardians, once they have experienced a life-changing traumatic event involving ICTs, remain suspicious of all technologies and their ability to cause harm. No matter the next latest and greatest innovation, Enthusiasts will be interested in trying it and Guardians will be cautious and limited in their use. Technologies, no matter how innovative, cannot change a person's user type once it has been established. All further technological experiences are filtered through the perspective of their type. There is no technological innovation that will change a Guardian into an Enthusiast.

A person's ICT user type can be viewed as a life trajectory. Life trajectories represent a series of events that a person experiences in their lives which, taken together, represent a pathway through that person's life (Elder & Giele, 2009; Fry, 2003; Giele & Elder, 1998). A person's individual path, or trajectory, is influenced by the life events they experience and their interpretation of these life events. These occurrences, and the meanings they develop for a person, can dramatically alter a trajectory (Giele & Elder, 1998). If we view a person's user type as such a trajectory, the events in childhood (such as exposure to ICTs, positive or negative experiences, and encouragement to tinker) are the starting point of that trajectory. For Enthusiasts, this starting point represents a positive, tinkering, mentored relationship with technology, setting them on a lifelong trajectory of loving ICTs. This love often leads them into technologically focused careers.

Guardians, however, somewhere along their user type trajectory, experience a technological life event that is extremely disruptive and involves significant loss. For Guardians, a traumatic event involving technologies becomes the impetus for creating their sensitivity to the negative consequences of technology. Be that loss of a family (Natalie) or a job (Margaret), this traumatic event leads Guardians to have a much different trajectory compared to Enthusiasts: Guardians come to distrust and have a lifelong suspicion of technology. Their development into Guardians influences their life choices as well: they choose to limit their technology use and are cautious about allowing too many ICTs into their daily lives.

Traditionalists within any generation can only be observed as that generation ages. As young people, they are involved in many of the technologies readily available. They love technologies when they are young – much like Enthusiasts. However, over time, Traditionalists' love of technology does not move onto the latest innovation, as Enthusiasts' does. Instead, Traditionalists continue to heavily use the ICTs of their youth, rejecting newer ICTs.[1] Therefore, a given generation's Traditionalist becomes apparent as the generation members age and technology advances.

Other types are reinforced by their environmental surroundings over time. Socializers tend to have large families (with many children) and build large intergenerational social networks. They prioritize maintaining contact with these large networks, for which ICT use is incredibly important. As they age, they begin to mimic the communication patterns and devices used by their youngest contacts. While they are excited by technology and consistently find social ICTs fun, what really matters is how they can use a technology to maintain connections with others. An important part of their identity as Socializers develops throughout their lifetimes, particularly when they, as older adults, engage younger individuals, mimicking the young's communication patterns.

But what about younger generations? The ICT User Typology predicts that those who are young now – Millennials and Generation Z – will eventually develop the five user types. The common thought in popular literature on younger individuals (particularly Millennials) is that they are digital natives – they grew up using and surrounded by technology and, therefore, are innately able to use software, applications, and new devices. However, research has shown that younger individuals are not necessarily digital wizards: there is great diversity in

[1]Future research will need to determine how these differences between the Traditionalist and Enthusiast types develop. One important difference between these types is that Enthusiasts report having had technological mentors and being encouraged to tinker in childhood; Traditionalists report neither.

the ICT use and skills among Millennials (Helsper & Enyon, 2010; Weiler, 2005). Rather, like all generations before them, Millennials simply have the advantage of society considering them the legitimate users of recent technology because they are young (Larsen, 1993). When Millennials have been replaced by newer and younger generations and technology has advanced, they will no longer benefit from these stereotypes.

But why do we not see Traditionalists or Guardians among Millennials? Currently, it would be difficult to spot Traditionalists and Guardians among Millennials because the Traditionalist and Guardian types develop over the life course. Millennial Traditionalists blend in with Enthusiasts because both types love the technologies of their youth and early adulthood – and Millennials are young! After all, Traditionalists do not reject all technology, they just reject technologies that are introduced after they reach mid-adulthood.[2] Until a generation reaches mid-adulthood, Traditionalists blend in with Enthusiasts, both sharing their love of the present technology. Guardians have had a traumatic event by mid-adulthood that they credit with changing their technological perceptions. Millennials, as a generation, have not yet reached mid-adulthood. Many who will eventually become Guardians have not yet accumulated such a traumatic life event.

Socializers, Enthusiasts, and Practicalists may be more easily spotted in the earlier adult years of a generation. Practicalists express that they have a lifelong appreciation for ICTs as tools, and in younger generations they express that they see ICTs as being important for specific tasks, or that they use them in specific areas of their lives. Enthusiasts will express a love for all ICTs and demonstrate heavy use across many areas of their lives. However, because it will be difficult to separate Enthusiasts from Traditionalists in a younger generation due to their shared love of technology, it is important to examine other hallmarks of the Enthusiast type such as technological mentorship and tinkering experience. Socializers may be spotted because of their focus on maintaining connection – but it will be important to remember that they will not be speaking about how they need to communicate with younger individuals and mimic their communication patterns – because they are the younger generation!

Despite these user types developing over time, it is important to note that these types are *not* just specific to older adulthood or a single generation. The next sections focus on exploring each user type through the secondary data available from generations other than the Lucky Few. These data come from the other four generations that are present as adults in the United States Society: the Good Warriors/WWII Generation (born in 1909–1928), the Baby Boomers/ Boomers (born in mid-1946–1964, one of the largest generations ever born in

[2]Much of the discussion about digital natives assumes that technology will not progress or develop beyond our current abilities. Technology 50 years from now is likely to have as drastically changed as it has over the past 50 years.

the US), Generation X/ Gen Xers (born in 1965–1982, a small generation), and Millennials (born in 1983–2001) (Carlson, 2009).[3] The secondary data presented in this chapter suggest the universality of these types among several generations, countering the idea that all younger people are Enthusiasts.

The secondary participant data represent the analysis of 22 individuals who were friends, family members, and coworkers of the older adult members of the Lucky Few generation, the focus of this book. These individuals underwent a single interview with the researcher, either in-person or over the telephone. The secondary participant data are weighted more heavily toward Boomers (as the vast majority of the friends, cousins, siblings, and children of the Lucky Few generation were Boomers), a few Gen Xers, and a Millennial (all of whom were children of the Lucky Few participants). Since this chapter is specifically focusing on user types in generations *other* than the Lucky Few, data from the friends, family members, and coworkers of the primary participants also from the Lucky Few generation are not presented.

The 22 individuals analyzed in this chapter, their generation and their user type are illustrated in Table 2.

The Boomer generation is so large that scholars have suggested splitting it into two halves for analysis. While the late and early Boomers have more in common with each other than other generations before (WWII Generation) and after (Gen Xers), such a split can make the data easier to conceptualize (Reisenwitz & Iyer, 2007), so the data in this chapter separate Boomers into two such groups: early Boomers (born in mid-1946–1954) and late Boomers (born in 1955–1964).

Enthusiasts

Enthusiasts see technology as a fun toy to be played with and are excited to incorporate technology into every aspect of their lives. Enthusiasts were observed in the WWII Generation, Early Boomers, Late Boomers, and Generation Xers.

Bob, a WWII Generation member, was an Enthusiast who spoke at length for his love of technology, as Enthusiasts tend to do. When asked to define technology, he spoke about its long history, then turned toward how he used the computer:

> Well, [ICT] could be anything from the radio in the past or you could go back to dot dot dot (Telegraph) and all the way up to the Internet. It's the Internet today. And everything in between [...] I use it for a lot of different things. We have a mail-order

[3]Generation Z individuals are just starting to enter adulthood, with the eldest members born in 2002. Since generations are typically approximately 20 years in length, and often are bookended by major historical events, it is unclear if Generation Z has ended or will continue for several more years.

Table 2. Secondary Participants by Generation and ICT User Type.

		ICT User Type				
		Enthusiasts	**Practicalists**	**Socializers**	**Traditionalists**	**Guardians**
Generation/Birth Cohort	Good Warriors/WWII	Bob				
	Boomers	Lauren	Bobbie	Kate	Bette	Amanda
		Peggy	Charles	Julie	Marge	Marcy
			Dilly		Veronica	Mya
			Donna			
			Erica			
			Tom			
	Gen X	Adriane	Allison	Lynn		
		Chloe				
	Millennials		Katrina			

> business and we have a website, and a catalog and I always put together the stuff for both of them. And I do a lot of creative things too. I check my email. I've gotten so I look at things like Al Jazeera, foreign news places, and that's essentially it I guess. I'll look at what different people are saying about different things. (Bob, 1928, WWII Generation)

Bob is enthusiastic about technology and speaks about all the different ways he uses the internet: to run his business, to communicate, to read the news, to read comment forums. As an Enthusiast, Bob also speaks about how he stretches his use of the computer and internet across all domains of his life: he uses them in his work, leisure, and family. (This is an excerpt of a much longer quote in which he speaks about all the ways he uses his computer and cell phone.)

Many Boomers were also Enthusiasts. Peggy, a friend and former coworker of Belinda (herself a Practicalist), discussed her love of technology:

> I just like technology. I like gadgets, I'm a gadget person. I have a desktop computer in my home. I have a laptop that I use at work with a docking station with a larger monitor and a keyboard and a mouse, but I can take the laptop with me if I need to do presentations and stuff. I just received a new iPad, third generation. I have an iPhone and I have a Blackberry, I have a digital camera, trying to think of what else I have, and I tend to use all of them. Oh, I have a flip camera too, I use them all. I mean I have two phones because the iPhone is my personal and the Blackberry is my work. So that's the only thing that I actually have redundancy, but I need to use both for different reasons. I also have the netbook, I'm not using the netbook as much because the iPad is replacing the netbook. (Peggy, 1947, early Boomer)

Peggy, like all Enthusiasts, describes her use of ICTs with passion. She describes her many "gadgets" and the way that she uses them in an almost breathless fashion, reminiscent of Fred (Chapter 2). It is not uncommon for Enthusiasts to describe their technologies as 'gadgets' in a loving way. Such phrasing brings out the play and fun aspects that Enthusiasts experience when experimenting with technology. Enthusiasts, be they WII Generation, Lucky Few, Boomers, or Gen Xers, all speak about technology in such excited ways. Lauren (a late Boomer), shares:

> These children are spoiled; they have no idea how simple they have it. Technology makes life so simple whereas we didn't come up like that. We didn't have that. These children don't know how simple this life is as far as this technology. It makes life simple, so

> simple. You can pick up your cell phone or you can email any person you need to. Email is like super quick especially if a person is always on their computer checking their email. If they have access to a computer. The phone, super quick, you can have them pick up or you can leave a message. It's simplified, it's super convenient, I would put super in front of convenient and so simple. All the technology makes life so simple […] You can now alleviate paper with portable scanner that you can have in your own home. You can have a fax. It's really awesome. I love all this. This technology makes life so simple. We thought it was a big thing when we got a remote for the television as kids, that was totally awesome. I'm telling you when the electric typewriter came around I was in awe with that and the remote control. [When I first] saw the cell phone and I thought 'I don't believe this can work' and it did. It was big and blocky, not small and fancy like now. With technology you can communicate anything anywhere via Skype, that's awesome. I have friends in China we spoke every Sunday morning with my son before he went away to school. Anything you need to do. Gwen and I should get more into Facebook. I think we should get more into the email thing. What else is there? Oh, Skype. We should do Skype. In the future one of us may move away, or if we go on vacation I might Skype with her. (Lauren, 1960, late Boomer)

Like Fred, Lauren states that she *loves* technology. It is "awesome," she "loves it," it makes life "simple." Most importantly, with Lauren's quote one can see that she has a typical Enthusiast's lifelong technological love affair. She reminisces about how she has been impressed with technology since a child, speaks about all the different uses for ICTs in her life, and how she uses them across her family, work, leisure, and community contexts. Not only does she use many ICTs, there is even more she wants to try. Her excitement about technology jumps off the page. Importantly, technology is a toy, something to be experimented and played with, as Peggy shares:

> So, I tend to use a lot of newer stuff, like right now I'm playing with SpringNote and Via Lock and EverNote […] So, I'm always kind of picking and choosing ones that will end up being some of my favorites that I'll continue and integrate into my ICT world. I'm always looking for new stuff, new innovative Web 2.0 tools to support teaching and learning. So, I think of all that kind of all blended together. (Peggy, 1947, early Boomer)

Peggy shares how she "plays" with technology, much like Fred, Alice, and Harry (the Lucky Few Enthusiasts) do. Play is an essential view that Enthusiasts incorporate into their experiences with ICTs; they feel that technology is a toy, a fun experience. Adriane shared her motivations in acquiring what she considered her new toy, at the time, a cell phone:

> When I got my cell phone, I remember it was one of those giant silver things that looked like a mini suitcase, I think I was in college, probably 1993 [...] It was just a new toy. I was 18 and wanted to be up in the latest technology, the latest trend. It wasn't so much that I needed to communicate with anybody, just something I had to have. (Adriane, 1974, Gen X)

Like all Enthusiasts, Adriane, a Gen Xer, shared her motivation in acquiring her cell phone was to have the latest "thing." Adriane had no specific use for her cell phone – she did not need it for a task or to connect to someone – she wanted the latest toy to play with. Love and toys were common language seen throughout the Enthusiasts of all ages.

Enthusiasts, like Chloe, are always looking for the latest "gadgets" to experiment with:

> I think it's more about looking for a new gadget, looking for the latest thing to use. I'm looking at what's out there on the market. (Chloe, 1969, Gen X)

Enthusiasts spanned many of the generations of secondary participants, including the WWII Generation, Boomers, and Gen Xers. Their love of technology and their view of ICTs as toys and play echoed their fellow Enthusiasts in the Lucky Few Generation. In addition to the generational diversity observed in Enthusiasts, there was also large a number of Practicalists among the secondary participants.

Practicalists

Unlike Enthusiasts, Practicalists do not view ICTs as fun toys to experiment with, but instead as tools that allow them to accomplish a specific task or purpose. Practicalists were observed among Boomers (both early and late), Gen Xers, and Millennials.

Bobbie, a friend of Nancy's and a fellow assisted living resident at the same facility, shared that she used the computer most notably to find information. To Bobbie, an early Boomer, the computer was one piece of her life, not the most important part of her life:

> I am computer literate. We have a computer available to us here that I use frequently for almost anything. It's amazing the technology, well we're talking about technology of the computer, I mean you can logon to just about anything and find it and if you can't there's something very wrong. I also like to read, do

> crossword puzzles, and just basically spend time with my family of course. I don't do any of the email, Facebook or all that. [I'd like to use] email. (Bobbie, 1951, early Boomer)

While Bobbie enjoys using the computer and would like to learn to use more functionalities (she also uses a cell phone), she views its use as only one among her list of activities, which include non-digital tasks such as reading, spending time with her family, and doing crossword puzzles. For Bobbie, the computer is primarily a leisure device. Similarly, Donna, another early Boomer, shared that she also used the computer for several activities:

> As a matter of fact, I got it on [the computer] right now. I play games a lot on it, but I do get on the internet to look up information and stuff. I just sent an email a couple days ago. (Donna, 1949, early Boomer)

Both Donna and Bobbie are Boomer Practicalists due to their views that the computer is one part of their life, a tool that allows them to accomplish some important tasks in their lives, but not a toy that's use they constantly seek to expand. Other early Boomers commented on how their use of ICTs was limited to certain functions and purposes. ICT use was just a piece of their lives, not a subsuming activity like it is for Enthusiasts:

> I don't use Twitter, I don't use Facebook. I don't have a need. If I was in the government setting that I needed to have a Facebook or Twitter account then I would certainly explore that as a need. Just because a technology is out there doesn't mean that's appropriate for use [...] As a guy who was [once] a computer weenie [geek] I have basically cast aside some of these new technologies because they just don't fit my need. It's not like there's not validity there [for people to use them]. (Tom, 1954, early Boomer)

Tom's use of the computer is extremely focused. He does not reject Facebook or Twitter because he dislikes them, but because he does not have a practical need for using them. He can easily think of applications where he would want to have such accounts (such as if he was involved in government), but currently does not need such ICTs. Such a practical, function-focused analysis of technology is a hallmark of Practicalists. Many Practicalists echoed these sentiments, discussing how the ICTs they chose to use matched their needs and purposes:

> [This nutrition site is] a website where you can track your nutrition, your calories, your vitamins [...] It's wonderful to see how many glasses of water you drank and if you're keeping up, but I don't think I want to sit and stare at the screen. I have a nice screen it's one of those wide ones. Even that I don't want to sit

> here in this uncomfortable chair and be there any much longer. I want to be out and be involved with life, I have a life outside of the computer. (Dilly, 1953, Boomer)

Dilly, in her discussion of her computer use, describes how she finds a nutrition website helpful in tracking calories and nutrition. While this website is a great tool for Dilly – it is only a tool. Like any tool, it has a purpose and also times when it is not useful. Hence she finds times to log off and leave her technology to do other tasks.

Late Boomers also shared their views on ICTs as tools. For instance, Erica (Boris' daughter) shared that she used her computer mostly for playing games (and used her work computer for work tasks). She had social media, but used it in a limited way to observe what her friends were doing and to participate in reunion activities:

> I'm on the computer quite a bit playing games. I play World of Warcraft, it's a multiplayer game where you have your own character that you build. I don't know if you're familiar with the game at all [...] I do everything on my cell phone. As far as my father, Boris, is concerned I really only call him on his home phone because even though he has his cell phone he never takes it with him anywhere. My father and I are both on Facebook, but we don't talk on Facebook at all. I very rarely post on Facebook. I use it just to kind of find out what's going on with everybody else and for class reunions and stuff like that. I don't do a lot of posting on it. (Erica, 1964, late boomer)

For Erica, her personal computer is predominately a leisure device. She has no intention or desire to spread its use to other areas of her life, nor is she eager to play with it to find new functions or applications. While some late Boomers spoke about ICTs in terms of the functions they used them for, some were very explicit about their views of technology as a tool. Charles, a late Boomer, shared his view on technology:

> Well, ICTs are tools. Tools to achieve a purpose. I don't look at anything else from a technology perspective other than that. I don't play video games. I'm as jaded as hell when it comes to technology. I see, I immediately begin to look for a purpose for those tools to be used because that's what they are. They're tools to achieve an endpoint, or a goal, or a project goal, or a purpose, or some kind of project development. I am still in that mode where I didn't grow up with a cell phone and I certainly don't need one. I work with a laptop, I love having one, it has been very good but I don't necessarily need one. I mean I've seen it all develop and it's just, to me they're just tools. That's the first thing

> that comes to mind. Technology comes to mind as a tool. (Charles, 1964, late Boomer)

In Charles' mind, ICTs are tools that serve a specific purpose: to reach goals, develop projects, or make something. Unlike an Enthusiast, he does not want every ICT he could possibly use, but rather is specific in what technologies he uses, what purposes he uses them for, and why. He prefers a laptop to a cell phone, as he has use for a laptop. The focus in Charles' statement is on how ICTs are task-specific tools that allow people to accomplish activities.

Younger secondary participants were also observed to be Practicalists. Allison, a Gen Xer, spoke about how she carefully delineated work ICTs (such as her work computer) from more personal uses of technology (such as Facebook and Twitter). In her work, as an administrative assistant in a front-facing office, she was careful to not use what she considered leisure ICTs (be they devices or software) that would make her appear unprofessional. While she would occasionally read and quickly reply to personal emails that were sent to her work email address, she did not look at social media or use her cell phone while at her desk. She elaborated further on how she felt that all ICTs were tools, and as tools different ICTs would appeal to different individuals:

> [When one speaks about ICTs], you have different populations that are using different things, some that only use Twitter, that only use Facebook, that only use their email accounts, and some that don't want to receive those things. Those people get information from watching the morning news or they listen to their radio as they're driving to work because they don't want to go on the computer, or they don't want to deal with the tweets. So, you have all of these different places trying to get their message across in order to make sure that their messages are received. (Allison, 1977, Gen X)

When reflecting on the uses of technology, Allison considered how institutions, companies, and other organizations use them as communication tools to reach the general population. In order to reach everyone, an organization would have to use different types of media as different individuals are using different media forms. Her view reflects that individuals use ICTs for specific purposes, just as her own use was targeted.

Katrina, a Millennial, also shared her Practicalist views. She compared her own ICT use to that of her father, Harry, an Enthusiast. She described her use as more limited than her father, stating that she mostly used her cell phone and computer to communicate, while she also used her radio to listen to music:

> The things that I use right now are my cell phone and my computer to communicate. That's primarily what I use. I listen to the radio [...] (Katrina, 1984, Millennial)

Later in our conversation, Katrina would speak about how Harry had influenced her ICT use, particularly when it came to using the digital editing software she used in her work. It is important to note that Katrina emphasized how much more technological skill her father had, despite being several generations older:

> Information technology gives us (my father Harry and I) something else to talk about and some common ground, so he talks about what's new, like the newest cell phone. I've shown him my computer, or my work, and we'll talk about it. Whereas my mom might not know what I'm talking about or might not care just because she doesn't know about it. For my job I edit video on a computer a lot. Harry has started to do that, and he'll have suggestions for me and so we talk about that a lot. I call him for questions more than not. He's sort of known as the computer guy for the family. My brother, my sisters, my aunts and uncles, they all call my dad for computer problems or computer suggestions. He is the technical support guy [for our family]. [But] they're not usually the oldest. (Katrina, 1984, Millennial)

There is much to unpack in this statement. First, Katrina stands in contrast to the stereotype of the young Millennial who is a digital native. Her daily work involves video editing and heavy computer use; however, she often approaches her father Harry, an older adult, with questions not only generally on computers, but also specifically on video editing. Second, we can see that not all young people are Enthusiasts but can vary in their approach to technology. Katrina approaches her ICT use as a Practicalist: she seeks to better understand the technology she finds a use for in her everyday life, such as video editing. Her father, an Enthusiast, picked up video editing as a side interest when his daughter mentioned it to him. He, as an amateur video editor (due to his Enthusiast perspective and interest in playing with new technologies), is often able to give his professional video editing daughter tips and advice. The contrast in these two approaches illustrates how Katrina is a task-driven Practicalist, but Harry a fun-seeking Enthusiast.

While Practicalists are function-focused ICT users, Socializers focus on the ability of technology to build, maintain, and grow relationships, as can be seen from several secondary participants.

Socializers

Socializers emphasize the social and relationship connections of ICTs. They see technology as a vital way of continuing relationships with others, as another way of connecting to their important family, friends, and community members. Socializers were observed among Boomers and Gen Xers.

Julie, a close friend of Alice (an Enthusiast), spoke about the power of ICTs to enable connection:

> There's so many ways to communicate now you know, it's like you can phone, you can text, you can write, you can email, and you still have the old-fashioned way—you can go down the street and knock on their door […] The lifestyle of modern day people of being miles away from each other is when the technology has really been helpful. It's like less distance between us because we can communicate. When I do get come to visit [from across the country] I sometimes feel like there's less of a gap between us […] Things feel more immediate, they don't feel as distant. I think technology has really shortened the concept consciously of having distance between people. I feel like being close to someone is enhanced by being able to text and email. If not for texting and email, I would only see Alice one month and I'd be back the next month. Instead we feel like there is a continuum between us because we text. It's like you didn't lose a step or continuity in the information and relationship. (Julie, 1953, early boomer)

Julie, an early Boomer, begins her discussion of ICTs by speaking about communication. ICTs can lessen "the distance" between people, thereby strengthening the connection and relationship between people who might not be located in the same area. The issues that Julie concerns herself with are mostly in maintaining and nurturing her relationships. Technology is not a tool, or a fun toy, but a way that Julie draws herself closer to her friends and family members.

Socializers, when asked to define ICTs, emphasize the communicative aspect of technologies. When Kate, also an early Boomer, speaks about how she used work-based ICTs prior to her retirement, she focuses on communication technologies such as video conferencing, rather than the use of software such as word processing:

> I have a cousin in Milan and every day she writes what's going on in her studies and I read her blog every day. Blogs and Facebook I read once a day, I text, I'm not big on Skyping but have Skyped. While I was working did many technology-driven meetings with video cameras and with the screen on the other side […] Video conferencing that's the word. What's my definition of information and communication technology? The ability to communicate, the ability to transfer ideas from one person to another using a different media than face-to-face. (Kate, 1954, early Boomer)

For Socializers, like Kate, the focus of ICT use is on communicating with others. All the ICTs she mentions: blogs, Facebook, texting, Skype, and video conferencing all have communication embedded at their heart. Kate went on to share how she specifically uses technologies in her relationships with older individuals, such as her cousin, Mary:

> The speed at which I have technology between myself and older people is not necessarily based on the information as much as it is the mentoring and the conversation and the learning. (Kate, 1954, early Boomer)

The use of technology between Kate and those older than her is for "conversation," "learning," and "mentoring." These words suggest the importance of the relationship and the role of communication in strengthening the relationship from one of just familial or friendship bonds, to one of mentorship. Technology is not an information source; it is a way of enriching social bonds. It is a way she can connect to the older generation – a way she can learn from them – not just through information, but through conversation and enrichment.

Lynn, a Gen Xer, also spoke about ICTs from a communicative perspective. While her background in education influences her, she focuses overwhelmingly on the connectivity of such technologies:

> When I think about technology, I think about social media, I think about information literacy skills. I guess my viewpoint is from an educational perspective. I definitely think that's the lens I look through. I view information and communication technologies as connection through technologies that provides channels of access and information. (Lynn, 1972, Gen X)

For Lynn, information technology is a way that people connect to one another. This focus on connection and socialization is a hallmark of Socializers.

As discussed earlier in this chapter, Socializers, Practicalists, and Enthusiasts are easier to spot before mid-adulthood than Traditionalists and Guardians, who are more easily identified later in the life course. Several Boomer Traditionalists, who love the ICTs of their youth, were spotted among secondary participants.

Traditionalists

Traditionalists love the technologies of their youth. As they grow older, however, they come to reject newer innovations. Traditionalists' lives are so full of the technologies they love; there is no room for modern innovations.

It is important to note that Traditionalists only become apparent in a given generation as that generation ages into middle to late adulthood. Since Traditionalists love the technology of their youth into middle age; a Traditionalist in a generation that has not yet passed beyond midlife is not easily recognized – as the technology they would reject has not yet been introduced.

Traditionalists were observed in both early and late Boomers. Marge, an early Boomer, made a specific distinction between early technologies that she used (such as the telephone, television, books, and magazines) and those that she did not (such as the cell phone and the computer):

> Ham radio operators always fascinated me. Back when I was young they could talk to someone on the other side of the world, where the average person couldn't do that. Not back then. Now you can because of cell phones. That's been a change in my lifetime. I would say that an information or communication technology is something that makes noise. Something that doesn't make noise isn't an information or communication technology. But a book is information. And it is passed. So not a book [...] No not a book I wouldn't say. Because a book can be passed and is information just like the magazines I read. It's information I'm getting from somewhere else, from someone else. I would say it's a technology. In an old-fashioned way—an older form. Before gadgetry. I don't know what else to call it except for before gadgetry. (Marge, 1947, early Boomer)

Like a Traditionalist, Marge speaks about being fascinated with the technology of her youth, in her case, HAM Radio operators. As all Traditionalists do, she has rejected newer ICT forms such as the cell phone (she uses hers only for emergencies), the computer, and the internet. Traditionalists were also found among late Boomers, including Veronica.

Veronica rejected most newer ICT forms. She could, with some help (typically from her son), use the computer to send and receive email. But, in her own words, she was not "very good" at it. She often had her son use email and Facebook for her, as is typical among Traditionalists. In speaking about her friendship with Gwen (a Lucky Few Socializer, featured in Chapter 4), she stated:

> Gwen hasn't sent me much on email. She'd say like hi or something like that. She had me try and send her something one time and I can't remember what it was but, [it didn't send]. Because I'm not very good with the computer myself so [...] like even on the internet, I don't like sending things like messages to my aunt in Texas on the internet with Facebook because I don't know how to do that stuff. I have to have my son do it for me. (Veronica, 1962, late Boomer)

Veronica serves as another example of why age is not the sole determination of skills or motivation to use advanced ICTs. Her friend Gwen, as a Socializer, was much more motivated and used many more ICT forms than Veronica, despite an almost two-decade age difference (Gwen being older). Veronica, as a Traditionalist, has rejected many of the more advanced ICT forms (computers and social media, and to a certain extent, her cell phone) and is instead an indirect ICT user (as many Traditionalists are) through her son. I asked Veronica if she would like to learn how to better use her computer or social media. She responded that she was not really interested in it. Like most Traditionalists, it is a lack of interest that leads to less developed technical skills; as opposed to a lack of technical skills leading to disinterest. Gwen indicated that she had often offered to help Veronica learn to use more technologies, but Veronica declined, always stating that she was not interested.

Bette also is an example of a late Boomer who is a Traditionalist. She is a heavy user of television, uses radio in the car, a stereo when with her boyfriend, and a cell phone. While she can text and uses a computer heavily for her work as an administrative assistant, she prefers to use older forms of ICTs from her youth in her free time:

> On television I watch all sorts of shows. I watch the news. Unless I'm in the car I don't really listen to the radio. My boyfriend has a great stereo, so I mean we'll listen to his tunes or tunes that I like. That's pretty much it for the radio. We'll listen to a basketball game; we just did that recently. I have a cell phone, and that's how I keep in touch with my mom [...] sometimes I'll search on the internet with it and get text messages. I still would rather call than text. (Bette, 1964, late Boomer)

Bette establishes that while she can text and can use a computer, she often chooses not to in her personal life. While she is required to have a smartphone for her work, and occasionally uses it to answer a work email or search the internet, she does not use it for personal use, besides the occasional phone call or text. Bette, however, loves television and uses it frequently. Like June, a Lucky Few Traditionalist (Chapter 5), Bette has a high level of exposure to computers in her work life. Outside of work, however, she has little motivation to use a computer. Her life is already full of more traditional forms of media.

While Traditionalists reject newer forms of ICTs in their lives, they love older forms. Guardians, however, restrict their use of all ICTs in their lives, regardless of their vintage.

Guardians

Guardians are suspicious and cautious around all ICTs. They have a general distrust of all technologies and heavily regulate their use. In particular, they believe that if ICTs are not used in a controlled fashion, they lead to the degradation of society. Several Boomer Guardians were identified among the secondary participants.

Mya, an early Boomer, like many Guardians, is particularly concerned about the impact of ICTs on younger people. In comments reminiscent of Margaret

(Chapter 6), she speaks about ICT use isolating individuals, when they should be interacting with one another:

> The whole technology thing it annoys me because they assume everybody is into that and has the means to get the stuff and brainpower to learn it all. It's annoying to me. I know younger children drive me crazy, including my family. We'll get together and instead of playing board games they all sit with their little DS or iPod or smart whatever and they don't play this game on their little thing together. Cell phones I believe have a very good purpose, I have one, but I get very annoyed. It promotes feelings of, 'I need to know right now, quick, I want it now, I need it now' instead of waiting for things. That's how I feel we really are getting to be as a society. It trickles down to kids wanting something right now, everybody is just in the moment. I think all this has a good purpose to a point, but I think it gets a little carried away. (Mya, 1948, early Boomer)

Mya discusses how technology has a purpose. However, instead of being used correctly (in a restricted and controlled way), it is often used inappropriately. Not used in moderation, technology subsumes and addicts its users. It leads to immoral behavior: isolation, want, impatience, and excess; all negative connotations of ICT use. Such use also degrades relationships. Instead of young people communicating and building their relationships face-to-face, they are devaluing these relationships by using digital devices.

In contrast to Socializers who view ICTs as connectors, Guardians view ICTs as separators. Marcy, an early Boomer, discussed how technology can cause people to withdraw, becoming reclusive:

> I know how wonderful computers and all this technology and information and everything can be, especially when you think of well medical breakthroughs and sharing information about people. But I do think that it's starting to make people very [...] very secluded, not getting in touch with people as much and [...] I mean there's a big difference between a text and actual being with a person and talking with them [...] I think [technology is] making a lot of people very reclusive. I think that technology makes people share too much about themselves sometimes. Then you hear of all the problems like identity theft. It seems like every night on the news [criminals are] coming up with new ways [to abuse] what people put on Facebook. (Marcy, 1950, early Boomer)

For Marcy, as all Guardians (regardless of generation), technology is seen as unlocking negative traits we all have inside of us. She differentiates between "being with a person" versus using technology. Technology also leads us to share too much information online, which opens people up to another threat of technology: identity theft. Like most Guardians, she is sensitive to the ways ICTs can be misused and impact our information security.

Marcy can identify that ICTs bring positives to our lives, such as medical breakthroughs and being able to share information. However, when ICTs are being misused (as most Guardians believe that they are), they lead to us down a path of immoral behavior and put our safety at risk.

Amanda, an early Boomer, used a cell phone and even had a Facebook account. In her view, like many Guardians, one *had* to use such technology:

> But for a lot of people [technology] just controls their life. I don't want to be controlled that way. I don't know if it's an obsessive-compulsive thing or if it's taken the place of alcohol. I don't know what the psychological thing there is, but I think some people are more intent on making a connection with people directly. And some people aren't. I go online and I'm on Facebook because of my grandkids so I can see that there's benefits to that [...] You almost have to [use computers] now, because there's so many people that are doing it, if you're not you're kind of out of the loop. I want that kind of contact [...] unless I want to be a hermit and never leave the house. But I don't. I want to stay connected with other individuals. (Amanda, 1947, early Boomer)

Guardians of any generation are deeply suspicious of technology and how it can be used. They personally restrict and carefully self-monitor their own use to avoid the moral pitfalls of technology, while understanding that such use is often required to be a participant in modern society. Amanda, for instance, had chosen to not use a smartphone, as she has observed people being "controlled" and addicted by their smart devices. However, she has both a cell phone and keeps a Facebook account. She uses Facebook to connect to her children and grandchildren, feeling it is a necessity in the modern world.

Following many Guardians, Amanda does not differentiate between "newer ICTs" and "older ones" in determining their value and their potential to cause harm. She shares that she has already unplugged and no longer watches her television, an older technology:

> I've already unplugged my television. There's a few times there's programs on there I want to see, but I forget when they are, and I forget to turn it back on. To plug it back in. For whatever it is. So, I just weaned myself off of it. (Amanda, 1947, early Boomer)

In the case of some ICTs, they are seen as so morally corrupting that a Guardian will reject them outright. Unlike Traditionalists, who value older ICTs above newer ones, Guardians make no such chronological distinction. Amanda, as a Guardian, like her older counterpart Jackie, has unplugged her television. The television, a device beloved by Traditionalists, is seen as contributing too much to moral decay by Guardians.

While no younger generations of Guardians were captured among the secondary participants, this finding is not surprising. Most of the secondary participants were either Lucky Fews or Boomers, leading to a higher representation of Boomers. (Lucky Few secondary participants were not featured in this chapter, given its exploration of user types in other generations.) The Guardian type likely develops over the life span and younger individuals are less likely to have had a traumatic life experience with a technology purely due to fewer years lived.

Different Generations, The Same Types

Overwhelmingly, the secondary participant data suggest that the ICT User Typology is not just a theory of older adult ICT use, but a theory that applies to ICT users in general. Enthusiasts, Practicalists, and Socializers are seen across the life span, while there is some evidence to suggest that Traditionalists and Guardians evolve over the life course.

Enthusiasts, Practicalists, Socializers, Traditionalists, and Guardians have likely always existed. They are not products of the computer age, but rather echo society members' approaches to technology throughout time. What changes with the march of ICTs and new innovations is the set of contemporary technologies in use at any given time. The development of a new technology, no matter how disruptive, will be embraced, rejected, or cautiously adopted depending on a person's user type. Technology does not change a person's user type; a person's user type changes how the person uses (and does not use) the technology. Exploring through empirical research how and when these user types develop (and if they can be influenced or changed during the life course) is the focus of Chapter 9.

Chapter 9

The ICT User Typology in Context: A Theoretical Perspective

The ICT User Typology is deeply embedded in several decades of gerontechnological research and a much larger field of technology use studies. This chapter focuses on exploring the relationships between the typology and existing gerontechnological research and other theories of technology perspectives and adoption. It starts by examining the connections between existing gerontechnological work on the factors thought to influence ICT use by older adults: most importantly demographics and gender. Next, it explores existing theories that are related to the typology, including the theory of IT Cultural Archetypes (Kaarst-Brown, 1995; Kaarst-Brown & Robey, 1999), the Diffusion of Innovations (E. M. Rogers, 1962, 2003; E. M. Rogers & Shoemaker, 1971), and the Taxonomy of Older Adult Digital Gamers (De Schutter & Malliet, 2014). Opportunities for further empirical investigation of the ICT User Typology are also highlighted.

The Gerontechnological Research and User Types

There has been an increasing focus in the gerontechnological literature on understanding how demographics such as income, socio-economic class, and education level impact older adult ICT use. Much of this literature has suggested that those who are of lower socio-economic class, income, and education levels are less likely to use modern forms of ICTs (Czaja et al., 2006; Friemel, 2016; González-Oñate, Fanjul-Peyró, & Cabezuelo-Lorenzo, 2015; Ihm & Hsieh, 2015; Peral-Peral et al., 2015; Van Volkom et al., 2014).

The ICT User Typology cuts across these class, income, educational, and work boundaries to suggest that a person's ICT user type is more predictive of their approach to technology than their demographics. Individuals in this study ranged from those in acute poverty (Gwen, Jackie, June, and Nancy) to the affluent (Charles, Mary, George, and Mindy Jean), with most of the participants in the middle class. Occupations ranged from blue collared positions to white collared, including electricians, IT workers, upper management, and the director/vice president level. Many of the women were in/had been in pink collared . professions, as was common for members of the Lucky Few Generation

(Carlson, 2008).[1] Participants ranged from those with General Equivalency Diplomas (or GEDs, an alternative to a traditional secondary school diploma) to two individuals with doctorates, with most participants holding a high school diploma, as was common for this generation (Carlson, 2008).

Based on these various demographics, there was no difference in a person's user type, with the exception of those who were Enthusiasts being more likely to choose IT careers. (It is likely that Enthusiasts' love of ICTs from early in their life influenced their career choice versus their career choice leading to their love of ICTs.)

Prior gerontechnological research has suggested a clustering of attitudes and uses of ICTs among older adults, providing hints at the user types. Studies have indicated that older adults have diverse motivations and reasonings for using ICTs including fun, usability, and usefulness (McMellon & Schiffman, 2002; Melenhorst, Rogers, & Bouwhuis, 2006). Such findings lend credence to the existence of the Enthusiast (fun) and Practicalist (function/usefulness) focused user types.

Other gerontechnological findings have suggested that older adults with larger social networks are more likely to use their cell phones than those with smaller ones (Petrovčič, Vehovar, & Dolnicar, 2016), suggesting that those with large social networks (like Socializers) have distinct ways of using communication technology to stay connected. Other studies have suggested that those older adults with higher levels of technophobia, or fears about technology, report more constrained, limited, and controlled levels of use (Nimrod, 2018). Evidence of such restricted use patterns lends support to the Guardian type's existence, as Guardians have the highest levels of technophobia and they strictly control and limit their ICT use.

While prior gerontechnological literature has hinted at the existence of the five user types there are other gerontechnological findings which contrast with the ICT User Typology. In particular, prior Gerontechnology work has suggested that gender is a driving factor in the adoption and use of ICTs. The ICT User Typology would suggest that it is not gender, but *perspectives* on ICT use that drive such behavior.

User Types and Gender

Much of the literature on older adults and ICT use has suggested that women are less likely than men to be using advanced ICTs (Kim, Lee, Christensen, & Merighi, 2017), experience greater ICT anxiety (Czaja et al., 2006), and report lower levels of self-knowledge (Helsper, 2010). Gender has also been shown to be important in many age-diverse domestication studies: women and men have different conceptualizations of ICTs (Ang, 1994; Cockburn, 1994; Habib & Cornford, 2002; Singh, 2001), both in terms of how they should be used

[1]In the Lucky Few generation women made tremendous strides in terms of workplace participation, but mostly in pink collared positions, such as administrative assistants, nurses, and teachers (Carlson, 2008).

(Buse, 2009; Lie, 1996; Livingstone, 1994), but also in who is seen as legitimate technology users (Cockburn, 1994).

Women were found across all five of the ICT user types discovered; however, men were only represented in the Enthusiast, Practicalist, and Guardian types. There is an open question if certain types are more gendered than others, particularly Socializers and Traditionalists.

All the Socializers were women. Such findings make sense: older adult women are much more likely to be involved in socialization activities and use ICTs to socialize and maintain bonds than men (Kim et al., 2017; Waldron, Gitelson, & Kelley, 2005). Certainly, age-diverse domestication research has suggested that men tend to view ICTs more as leisure devices for individual use and women view them more as devices of connection (Livingstone, 1994), suggesting a gender influence on the meanings that individuals give to technologies. While these prior findings support the Socializer type as being more feminine, other research suggests that there may be male Socializers as well. While not directly indicating the Socializer type, research has shown that older adults, regardless of gender, who have the largest social networks report highest use of their cell phones (Petrovčič et al., 2016). Given the influence that large intergenerational networks have on Socializers, such evidence suggests that it is possible male Socializers exist as well. After all, extroversion and joy in social activities are not a solely feminine trait.

The researcher's encounters with potential participants suggests that there are indeed male Traditionalists. Margaret's live-in partner refused to be a part of the study, because he, according to Margaret "isn't a computer user and isn't interested in any [technology] but television." Margaret reported that he loved watching television, listening to the radio, and using the telephone, but had no interest in learning how to use the computer, even though she often offered to teach him. (His love of the television often caused conflict in their relationship (Chapter 6), as Margaret, a Guardian, restricted her own television use.) Charles, a secondary participant and Margaret's neighbor, shared "Margaret's partner just said 'No. I don't want any part in that. Margaret can do it, but I won't.' So Margaret and I let the subject of your study drop rather than pushing him to participate." Such use patterns suggest that Margaret's partner was indeed a Traditionalist, who preferred the technology of his youth, but rejected later innovations.

There may be a tendency for older generations (such as the Lucky Few) to have more female Socializers and more male Enthusiasts than younger generations (such as Generation Z), as socialization has traditionally been seen as more "feminine" and technology as more "masculine." Instead of gender directly causing these differential gerontechnological findings, it is possible that a person's gender instead interacts with societal influences (such as societal expectations) to result in an individual's user type.

The ICT User Typology suggests that women do not have a single way of approaching or viewing technology, but five different approaches. While women may be more commonly found in one type or another, such as Socializers, this single user type or perspective on technology does not represent all women. This leads to an important question as to why gendered impacts are often observed in previous older adult studies. While older studies have suggested women have higher computer anxiety than men (Czaja et al., 2006; Laguna & Babcock, 1997), newer studies have demonstrated it is not gender, but attitudes toward ICTs that impact such feelings (Nimrod, 2018). These more recent findings lend credence to the fact that perspectives on ICT use have more influence on use than gender.

Older studies, by their nature, are examining a "different" older adult population than newer studies. Those who are considered "older adults" by a strict age definition represent a population that changes from year to year as "new" individuals, just turned age 65 (or the basic age set by the researcher), meet the requirements of inclusion (Birkland & Kaarst-Brown, 2010). This may account for differences in older versus newer gerontechnological studies' findings: the older adult population has changed.

It will be important to investigate if some user types are gendered or if they are more frequently gendered in some generations than others. While there are specific and interesting questions about gender and user type inspired by the Gerontechnology literature that must be explored further, the ICT User Typology can be integrated with other theories. The typology's integration with other theories of technology perspectives and adoption allows us a better understanding of how people approach and apply meanings to technology.

Integration of the User Typology with Other Theories

The ICT User Typology is the first theory that has concentrated on theorizing everyday older adult ICT use based on the meaning of ICTs in their lives. However, there are several other theories which have attempted to understand and predict ICT perspectives or technology adoption in a similar way to the typology, including Kaarst-Brown's (1995) IT Cultural Archetypes and Rogers (1962, 2003) Diffusion of Innovations. A third theory, the Taxonomy of Game Players (De Shutter & Malliet, 2014), suggests a clustering of five attitudes toward digital games among older adult gamers. All these theories, interestingly enough, indicate that individuals can be separated into five categories. However, these theories also differ in their application and understanding of technology use/adoption from the ICT User Typology. The following sections address the similarities, differences, and synergies between these three theories and the ICT User Typology.

Kaarst-Brown's (1995) IT Cultural Archetypes

Kaarst-Brown's (1995) IT Cultural Archetypes were developed from an intensive ethnographic comparative case study of two different North American organizations. Using data from over 80 interviews with these two organizations' employees, she discovered five emergent cultural views toward Information Technology (IT) that could exist in organizations. These five cultural views, termed archetypes, represent distinct views of IT, derived from people's basic technological assumptions (Kaarst-Brown, 1995; Kaarst-Brown & Robey, 1999):

- In a Revered IT Culture, IT is highly valued and looked to as a solution to many problems.
- In an Integrated IT Culture, IT is evaluated based on its match with user needs.
- In a Demystified IT Culture, IT is seen as a resource that any employee can use.
- In a Controlled IT Culture, decisions about IT innovation are believed to be a responsibility of organizational leaders.
- In a Fearful IT Culture, IT and innovations are seen as potentially harmful.

One can view these five archetypes as five potential subcultures that can exist in any organization, with each subculture having a distinct view on how IT should be used organizationally. This includes for what purposes and how IT professionals should be treated (Kaarst-Brown, 1995; Kaarst-Brown & Robey, 1999).

While IT Cultural Archetypes (Kaarst-Brown, 1995) focuses on IT cultures in organizations, the ICT User Typology concentrates on individual perspectives and meanings of ICT use. Despite these differences in what is being measured (subcultures versus individual perspectives) and predicted (organizational strategies versus ICT use), there are remarkable similarities between cultural archetypes and user types. It is likely that these two theories are capturing different parts in a chain of phenomena from the individual creation of ICT user types to the development of organizational IT subcultures.

Enthusiasts, who "love" and "cherish" their ICTs, closely match the attitudes observed in the Revered IT cultural pattern. In a Revered IT Culture, IT is honored and respected, and turned to for solutions for a wide range of problems (Kaarst-Brown, 1995; Kaarst-Brown & Robey, 1999). Similarly, Enthusiasts turn toward technological solutions in both their personal and professional lives, have great admiration for ICTs, and heavily advocate for greater use of ICTs in their workplaces. Enthusiasts enjoy working in technical environments and being involved in Revered IT Cultures. Fred was full of praise for his former employer (a secondary school) as he had been given free rein to use, invest in, and explore ICTs (a revered IT culture). When Enthusiasts encounter workplaces that do not value ICTs to the great extent that they do, they can be highly critical. Alice felt that the home healthcare agency she worked for was not using technology for the greatest benefit and was often suggesting ways to incorporate more ICTs into their workflow to her supervisor.

Practicalists, who see ICTs as "tools" that serve a specific purpose in their lives, correlate closely to the Integrated IT Cultural pattern. In Integrated IT Cultures, use by organizations is seen as a balance between technical capabilities and user needs. New innovations are evaluated on their contributions to an organization's well-being (Kaarst-Brown, 1995; Kaarst-Brown & Robey, 1999).

This focus on functionality and practical purpose in the Integrated Culture is remarkably similar to Practicalists' own focus on usability and function. In an Integrated IT Culture, those technologies which have greatest value are those that are determined to be the best solution (Kaarst-Brown, 1995; Kaarst-Brown & Robey, 1999); just as Practicalists place the highest value on ICTs that are useful. Practicalists enjoy working in Integrated IT Cultures, where the values and purpose of devices are clearly defined, and their use of helpful innovations is supported. Belinda was grateful for having colleagues that helped her identify useful technologies and a workplace that supported her through technical training and assistance (an Integrated IT Culture).

Socializers, who view ICTs as "connectors," want to use the technologies that younger generations are using. Socializers most closely mirror the Demystified IT Culture where users attempt to mimic the skills of IT professionals themselves: anyone can use IT, not just IT specialists (Kaarst-Brown, 1995; Kaarst-Brown & Robey, 1999). Socializers, who are heavily influenced in their ICT use by the younger generations around them, believe that these younger users are highly knowledgeable ICT experts, whom they wish to emulate and mimic. Socializers adapt to and adopt the communication patterns of their youngest contacts; learning text speak, choosing to use the same devices and applications, and constantly observing the latest technologies being used. Those in a Demystified Culture seek to be self-sufficient ICT users (Kaarst-Brown, 1995; Kaarst-Brown & Robey, 1999), much like Socializers do. Gwen commented on how she often made up text acronyms and had learned text speak to communicate with her grandchildren – meeting them using the technology they were using in the way they were using it

Traditionalists, who "love" and cherish the ICTs of their youth but have little use for more modern forms of ICTs, closely reflect the Controlled IT Culture discovered by Kaarst-Brown (1995). In the Controlled IT Culture, it is believed that decisions about IT should come from the top of the organization, and use should be carefully contained and constructed (Kaarst-Brown, 1995; Kaarst-Brown & Robey, 1999). This mirrors Traditionalists' own behavior in which they control which ICTs they use (the ones of their youth) and the ones they do not (the more modern innovations). It is likely that Traditionalists, who have little interest in using modern ICTs (unless they are required to for work), see the responsibility for determining technological adoption in the workplace as simply some other person's task, and not their own. Of all of the five types, Traditionalists are not only the least likely to adopt a new ICT, but also the least likely to discover new ICTs on their own. They rely, instead, on family members or friends to introduce new technologies; reflecting this abdication of ICT decision-making responsibility to others. It is quite possible that if Kaarst-Brown (1995) had examined the personal lives of the employees at the organizations studied, she would have discovered the split between work (where Traditionalists use modern ICTs, if required) and home (where Traditionalists use only the ICTs of their youth). An example of this would be June, who, as a Traditionalist, had used more modern ICTs when required to do so by work, but in her personal life she very rarely used them.

Guardians, who view ICTs with suspicion and are wary of their potentially negative impacts on society, reflect Kaarst-Brown's (1995) Fearful IT cultural archetype. In a Fearful IT Culture, innovations are viewed as possibly harmful, and in particular, as "harm to people" (Kaarst-Brown & Robey, 1999, p. 121). This culture is marked by a high level of technological anxiety (Kaarst-Brown, 1995; Kaarst-Brown & Robey, 1999). These concerns echo Guardians' fears of the potentially negative impact of technology, and most importantly, how this technology can harm. Margaret, as a Guardian, experienced such a high level of anxiety over the increasing digitalization of her workplace that she disengaged from her job, eventually retiring much earlier than she originally planned.

The ICT User Typology and IT Cultural Archetypes (Kaarst-Brown, 1995; Kaarst-Brown & Robey, 1999) should be viewed as complimentary, not competing, theories. Both provide substantial evidence that there is a grouping of five clusters of attitudes toward ICTs; and these five clusters are remarkably similar. It is likely that the user types of the ICT User Typology represent individual attitudes and beliefs toward ICTs, while the cultural archetypes represent the organizational subcultures that can develop in the workplace based on the user types present.

Kaarst-Brown (2005) has written about how a leader's underlying assumptions about ICTs shape the organization's overall outlook on innovation, IT culture, and its willingness to adopt new innovations. Interesting questions remain, however, as to what influence individual user types may have on the development of IT cultures depending upon their concentration in an organization. For instance, is it only the user type of the person in charge that determines the IT culture for that part of the organization? Is it possible, if an overwhelming number of employees in a sector of the organization have a conflicting user type from the leadership of that sector, that the employees determine the IT culture, rather than the leader? For instance, if an organization is led by a Traditionalist, but has mainly Enthusiasts and Practicalists as employees, will the organization become a Controlled, Revered, or Integrated IT Culture – all possibilities of the different user types observed? Will subcultures of these three archetypes develop, and if so, will it result in conflict and resistance?

Using the ICT User Typology in integration with Kaarst-Brown's IT Cultural Archetypes (1995), it is possible to understand how an individual's own user type could potentially impact the culture of an organization, or how differences in user types across an organization could lead to IT cultural conflict. It is also possible to understand how an IT culture that is significantly different from a person's user type impacts that individual, particularly in terms of job satisfaction and retention. Margaret, a Guardian, worked in an office that had developed a Revered IT Culture. She quickly became dissatisfied with her work, disengaged, and retired years before she had originally planned. Alice, an Enthusiast, was highly critical of her organization's more Controlled IT Culture and believed that they should introduce more innovation; she left her organization shortly after our interviews concluded. Both Margaret and Alice were dissatisfied with their work culture, but for dramatically different reasons. (Technology having too high of a status for Margaret versus technology having too low of a status for Alice.) Both were highly valued employees according to their friends, family, and coworkers; both represented an organizational loss. Mismatch between their user type and organizational IT culture led to their eventual disengagement.

Given the similarity of the ICT User Typology and IT Cultural Archetypes (Kaarst-Brown, 1995; Kaarst-Brown & Robey, 1999), it is important to emphasize a few critical differences between the theories. The ICT User Typology is an individual level theory that examines domestication of ICTs in everyday life, whereas Kaarst-Brown's (1995) theory of IT Cultural Archetypes is a group-level theory that examines culture in organizations. In particular, the ICT User Typology is focused on explaining the diversity of older adult ICT use, while IT Cultural Archetypes are focused on how cultural attitudes toward innovations impact organizational IT strategy.

The close relationship of these two theories, however, provides more evidence that the user types of the ICT User Typology exist and are universal across generations. Kaarst-Brown (1995) developed her IT Cultural Archetypes theory in two large organizations from an age-diverse workforce. Although her study was completed approximately 20 years prior to the development of the ICT User Typology, her participants ranged in age from their early twenties through their seventies, representing individuals born from approximately 1915 to 1975. This meant that her participants represented the WWII Generation, Lucky Few, Boomer, and Generation X generations. These five cultural patterns were not limited to members in a single birth cohort/generation. This lends more evidence that the user types are not generationally specific, as discussed in Chapter 8.

E. M. Roger's (1962, 2003) Diffusion of Innovations

Another theory that presents five categories of users is E. M. Roger's (1962, 2003) Diffusion of Innovations, which seeks to understand how ideas become integrated into our societies. This theory posits that ideas and technologies undergo a process of adoption, and individuals can be separated into five predictable groups based upon their rate of adoption (fast to slow). These groups differ in their ages, financial, and social resources (E. M. Rogers, 1962, 2003; E. M. Rogers & Shoemaker, 1971):

- Innovators tend to adopt an innovation early, tend to be young in age, and have high financial and social resources.
- Early Adopters adopt an innovation shortly after innovators, tend to be younger in age, and tend to have high financial and social resources.
- Early Majority Adopters tend to be slower in adopting new innovations and have above average social status.
- Late Majority Adopters tend to adopt an innovation later than most individuals. They tend to have lower financial and social resources.
- Laggards resist adoption of new innovations, tend to be older in age, and tend to have low financial and social resources.

There are two trends seen in the five categories proposed by Diffusion of Innovations: resources and age. Those who are more likely to adopt a new innovation are likely to have higher resources and be younger (innovators and early adopters) compared to those categories that are likely to resist adopting a new innovation (late majority and laggards) (E. M. Rogers, 1962, 2003; E. M. Rogers & Shoemaker, 1971).

All of the participants discussed in this book in Chapters 2 through 6 were older individuals which Diffusion of Innovations (E. M. Rogers, 1962, 2003; E. M. Rogers & Shoemaker, 1971) predicts would be the least likely to adopt advanced ICTs. Yet a great diversity of ICT use and ownership was discovered: older adults were not all laggards. Age itself is not a solid predictor of adoption. There was also no relationship between resource level and user type. Instead, a wide variety of both financial and social resources was observed across the participants: several of the participants were quite impoverished, many were middle class, and a few were upper class.

The ICT User Typology suggests that technology adoption and use are not purely a function of finances, social resources, or age, but rather is highly related to an individual's fundamental beliefs and the meanings they find in ICTs.[2] Given the diversity of ICT use in the older adult population, the characterization of older adults and those with low social and financial resources as technological "laggards" is inaccurate at best, and ageist at worst. For example, we observed innovations diffuse from Harry (a Lucky Few generational member) to Katrina (a Millennial), the opposite of what Diffusion of Innovations would predict based on age. Resources also did not correlate with adoption. For example, Mindy Jean is a Traditionalist with a healthy social network and is the spouse of a retired executive who rejects advanced ICTs. Gwen is a Socializer who lives in low income housing with a large social network who embraces social media and texting. From the viewpoint of Diffusion of Innovations, Mindy Jean would be much more likely to adopt advanced ICTs than Gwen, due to Mindy Jean's much higher financial resources. However, Mindy Jean rejects modern innovations and Gwen embraces those that enhance communications. Put simply: it was not a lack of resources that prevented Mindy Jean from using these devices, but the fact that these devices had little meaning in her everyday life. Conversely, a scarcity of financial resources did not inhibit Gwen's adoption.

Although the ICT User Typology and Diffusion of Innovations both contain five categories, there is no clear mapping across the two theories. This is not to say that resources, both financial and technical support and encouragement, are not important – they are critical. Nancy, a Socializer, who desperately wanted to learn to text, was prevented from doing so due to her limited income and impairments. If a cell phone that accommodated both her abilities and her

---

[2]Prior gerontechnological research has supported that income and educational levels are not always predictive of ICT use; as individuals who have low income and education levels are often digital participants (Eynon & Helsper, 2010).

budget was available, she would likely be an avid texter. There is still power in understanding how resources, functional ability, and knowledge impact time of adoption and, therefore, Diffusion of Innovations can be a useful theory. It, however, can be complimented by other theories that examine perspectives on ICT meaning, such as the ICT User Typology.

The ICT User Typology is not simply a metric for understanding when or at what rate an ICT will be adopted, but provides an understanding of *how* an ICT will be used and *why* it has been adopted. It is not the technologies owned that make a person any particular user type, but rather the meaning these ICTs hold to an individual that does so. Seeing an ICT as a "toy" is what made Alice, Fred, and Harry Enthusiasts, not that they all owned cell phones. In fact, not all of the Enthusiasts owned the latest smartphones: while Alice and Fred both owned smartphones, Harry lacked good cellular data coverage at his rural home so he opted to have a simple cell phone. Similarly, among the Practicalists the prevailing thought of ICTs as a "tool" did not lead this group to adopt similar cell phone models. Belinda had a smartphone, Cleveland had a simple plan phone and Boris had a pay-as-you-go cell phone. What motivated adoption of these diverse models was the everyday tasks these Practicalists completed. While Belinda was a college-level educator, necessitating her to purchase a smartphone; Cleveland had recently retired and found that a simple phone plan suited his needs as he no longer needed to be in constant contact with his workplace. Boris used a simple pay-as-you-go phone, as it provided emergency service should he need it on a construction site. Practicalists are not Practicalists because they own a certain type of device, or adopted a certain device at a similar time. They are Practicalists because they are motivated to adopt new devices based on their usefulness, which stems from their shared belief that ICTs are tools.

What can be seen in the above examples is that for any innovation, categorizing individuals simply by the time of adoption or type of technology used leaves a critical gap in understanding *why* individuals are adopting an ICT. More importantly, we do not know *how* they are using the innovations they have adopted. For instance, Alice, an Enthusiast, had a smartphone, as did Belinda, a Practicalist. If we placed Alice and Belinda in the same category because they owned smartphones, we would miss important and critical information on *why* these women adopted these ICTs. Alice owned a smartphone *despite* working in an environment that did not encourage its use; Belinda owned a smartphone *because* her workplace required it. Appealing to Alice is the "fun" and "newness," appealing to Belinda is the "practical" and "function" – and these views impact how these women use their devices.

While Kaarst-Brown's (1995) IT Cultural Archetypes and E. M. Rogers' (1962, 2003) Diffusion of Innovations theories focus on the general population and their ICT perspectives and adoption, another theory, De Shutter and Malliet's (2014) Taxonomy of Older Adult Gamers, specifically focuses on older adults and their perspectives on digital gaming.

De Shutter and Malliet's (2014) Taxonomy of Older Adult Gamers

De Shutter and Malliet (2014) studied the gaming habits and perspectives of 35 older adults who played digital games at least once weekly. From their qualitative study, they determined that older adult digital gamers can be divided into five categories:

(1) Time Wasters view games as ways to fill time in a productive way, particularly for improving cognition.
(2) Freedom Fighters view games as a positive way that they can avoid other daily tasks in their lives.
(3) Compensators play games because they fulfill social and cognitive engagement needs that they are no longer able to fulfill in other ways due to functional declines.
(4) Value Seekers use games to learn about other hobbies or interests.
(5) Ludophiles are passionate gamers, individuals who love playing games and identify as gamers.

It is important to note that in the taxonomy that De Shutter and Malliet (2014) have proposed, all of the older adult participants analyzed were using digital games. This is a different population than the older adults studied for the ICT User Typology, which included many older adults who did not play games.[3] The Lucky Few individuals who played digital games in the study included Alice, Fred, Harry (Enthusiasts), Boris (Practicalist), and June (Traditionalist).

Alice identified most strongly as a "gamer" and most likely could be categorized as a Ludophile. Fred and Harry, although they did not consider themselves gamers, also found games fun like Ludophiles. Boris occasionally played games, viewing them as a way to be active and fill his time, much like a Time Waster. June's game playing is more difficult to categorize. June rarely used digital games; she would occasionally play solitaire on her computer every few weeks. June's level of game playing would not have met the selection criteria in De Shutter and Malliet's (2014) study, given that they required older adults to play digital games at least once a week. This is likely why June's digital gaming is difficult to categorize.

While at first it may seem that these two taxonomies are unrelated given the very different participants they were developed with and the behavior they describe, there are some interesting correlations. First, De Shutter and Malliet (2014) suggest that these gaming categories differ on their views of gaming as a more pleasurable versus a more useful task – a similar continuum to the split between the fun-loving Enthusiasts and the useful-focused Practicalists. Among older adults, there is some type of continuum in perspectives as to if technology

[3]In particular, Guardians reject digital gaming, typically viewing it as a waste of time and a removal from the "real" face-to-face relationships they value.

is meant to be more fun or more functional. Second, De Shutter's and Malliet's (2014) Compensators lend support for Socializers' innovative use of ICTs for social interactions. Compensators view games as way to adjust for declines in functional ability (cognitive or physical) in order to remain active and socialize. Nancy, as a Socializer, viewed virtual bowling as a way for older adults to compensate for physical declines and socialize, although she herself did not use the gaming system.

While the Gaming Taxonomy and the ICT User Typology are not describing a similar phenomenon, there is a similar synergy, lending credence to the fact that older adults differ on the meanings they assign to ICTs – be they gaming technologies or technologies more broadly.

Opportunities for Further Research

The ICT User Typology opens a potential field of research into exploring ICT use, meaning, introduction, and display from the perspective of the five user types. Opportunities exist for critically validating the types and for expanding the theory.

Expansion of the ICT User Typology Beyond the Lucky Few (and Beyond the United States)

One of the largest potential areas for development of the ICT User Typology is expanding the theory beyond older adults and the borders of the United States. The method chosen, dialogical interpretive interactionist (Denzin, 2001) case studies (Yin, 2009), allowed a depth of data collection that was necessary to generate the Typology. For practical purposes, the intensiveness of this method limited the number of comparative cases that could be conducted. (Methodological choices are discussed more fully in Chapter 11, "Discovery of the ICT User Typology".)

The consequence of such an in-depth study is that it tells us little about the demographics of the typology. It is unknown how widespread these user types are in the general population. For instance, it is unclear what percentage of the Lucky Few generation is any particular user type. There is also little information as to if this theory is applicable across cultural boundaries or if the percentages of user types differ from generation to generation or from culture to culture. This, of course, is an opportunity to explore if the ICT User Typology is indeed applicable cross-culturally and cross-generationally.

Currently, work is underway to develop a survey instrument that accurately captures the five user types, allowing further testing both in the United States (where this study originated) and in other nations and cultures. Such a survey will also allow for further validation of the user types, while allowing us to understand interactions of the types with other theoretical perspectives currently being developed in Gerontechnology (such as current research on income, self-efficacy, and resource accessibility) (Czaja, Boot, Charness, Rogers, & Sharit,

2018; Friemel, 2016; Pick et al., 2015). Further research must also examine how these types develop – and if they can be changed or influenced.

Exploring the Development of ICT User Types

From the study of the Lucky Few participants, it appears ICT user types begin developing in childhood. Early interactions with technology have a large impact on the development of a person's lifetime technological perspective. Events through the life course to mid-age continue to shape a person's user type. These insights as to when types develop, however, are retrospective in nature. Retrospective studies are tainted by the fact that our memories are often not perfect and that the events we remember from long ago do not necessarily reflect the reality of what factually happened (Scott & Alwin, 1998).

Retrospective memories can be shaped by experiences that happen after the event in question, colored by our own perceptions, and at best, can often be blurry (Scott & Alwin, 1998). In some cases, this was proven to be true for the data presented in Chapters 2 through 6 – most people could only name rough time spans in which they started using an ICT (often half decade periods) and sometimes could not remember their original motivations for using a technology, particularly if such use first occurred in early adulthood or childhood. (For some older adults this was nearly 50–70 years ago). For the vast majority of participants, more specific dates and feelings could be recalled in a second interview after the participant considered the incident during the interval. However, even with improved recall of dates and feelings, such thoughts are being translated and filtered through their following life experiences and their current meanings.

Within the life course literature, such retrospective studies are common because of the difficulty in designing and implementing prospective studies (Elder, 1985; Elder & Giele, 2009; Scott & Alwin, 1998). Many have argued that when studying a person's current state, views, and meanings, those memories which are the most salient are the most important to having shaped a person; no matter how inaccurate. When it comes to understanding meanings and personal stories, it is often less important what factually happened to us as individuals, but more important what we believed happened: our own personal narrative of the events that shape our lives (Denzin, 2001). There is a power in those memories that we hold as creating an integral part of our identity, even if they are somewhat factually incorrect.

Despite the legitimacy of the retrospective studies in gerontology, the proposed genesis of these user types calls for longitudinal prospective investigation. Such studies could determine which childhood events are critical: the importance of mentors, the role of technological "tinkering," and the importance of shared family media experiences. Most importantly, prospective studies could determine *when* such user types develop and the factors important in their development. They could also determine if these user types are stable or if they change over the lifespan (and most critically, how and what influences these changes). To truly understand the development of these user types would require a multi-generational study of individuals, tracked from early childhood until older adulthood. Such a study, while likely to yield incredibly important data, would span the work of several generations of scholars.

The benefits of such a study, however, would be enormous. Such work could track how these user types develop and possibly change over the life course, but also provide an in-depth understanding of how an individual's user type interacts with a person's entire life trajectory. This includes how a person's user type impacts their work trajectory (chosen career, career path, time of retirement) and interpersonal interactions and relationship trajectories (intergenerational relationships, friendships, and family relationships).

Given the often-cited need for larger numbers of Science, Technology, Engineering, and Math (STEM) career holders in the United States (Gonzalez & Kuenzi, 2012; Kuenzi, 2008) and Europe (Directorate-General for Internal Policies, 2015; Microsoft, 2017), there are also important questions as to whether a person's ICT user type can be influenced, leading to more Enthusiasts and/or Practicalists in a given cohort. (Enthusiasts and Practicalists appear to be more likely to choose STEM careers than other types.)

While such future empirical studies should address if user types can indeed be changed or influenced, the ICT User Typology has immediate practical implications for practitioners, designers, and advertisers who are interested in the older adult population. Strategies for applying the ICT User Typology are explored in Chapter 10.

Chapter 10

Breaking the Digital Divide

While the ICT User Typology is a theory of technology use, it has practical implications for how to meet the needs and wants of an aging society. Split into four sections, this chapter provides (1) guidance to practitioners who work with older adults, (2) methods and considerations for overcoming the barriers to older adult participation in our digital society, (3) strategies for designers and advertisers to maximize the lessons from the typology to reach a growing graying market, and (4) lessons for childhood educational programs to impact the development of user types in childhood. This chapter is designed for a diverse audience and is meant to appeal to those who work in aging, information and library science, technology design and development, advertising, and education.

Practitioner Opportunities: Understanding and Supporting Older Adults

Many practitioners who work closely with older adults have noticed the great diversity in ICT use by their patrons, clients, patients, and customers; while also recognizing the potential benefits of digitalization (Freeman, 2005). However, it is often difficult for those working with older adults to verbalize this diversity and, more importantly, to understand how this diversity impacts older adults' access to digital services. Practitioners can use the ICT User Typology framework to customize services, opportunities, and living environments to maximize life satisfaction for each of the user types and their ideal relationship with technology.

Enthusiasts are the most technologically capable of all the user types and, therefore, the most likely of your clients to seek information online or fill out online forms. ICTs are a fundamental part of Enthusiasts' lives and inability to use them negatively impacts their life satisfaction. Given the centrality of technology to their identity, in residential/ managed care settings or when working with Enthusiasts with disabilities or age-related declines, it is important to facilitate access to appropriate and useable ICTs. Enthusiasts prefer to have their environments filled with technologies and are the most likely to accept (and even embrace) ubiquitous computing, including smart home innovations.[1] They have

[1]For a review of smart home technologies for older adults, please see Majumder et al. (2017).

high expectations for organizations they interact with to be technologically innovative.

Practicalists tend to have medium to high ability to use online services and resources. Diverse as group in their technical knowledge, their skills are often tied to the ICTs they use/used in their work. Those with high and diverse ICT use in their workplaces typically have the best skill set. Therefore, it is important to recognize that some Practicalists may need additional help in completing computer-based tasks, particularly if they were in a less technologically-focused work environment or position. Practicalists, unlike Enthusiasts, do not enjoy being constantly surrounded by technology and are more likely to prefer ICT-specific spaces in institutional settings that are task-defined, such as computer rooms and separate entertainment rooms.

Socializers have a high need for social interaction, and this flows into their media and ICT use. They also have a diversity of skills, with their highest levels of skills being in communication devices and services. They, therefore, may need assistance in using other forms of ICTs or filling out online forms. Socializers do not like using ICTs alone or using ICTs they consider isolating. Television watching, for instance, tends to be too solitary of an activity for most Socializers and they would greatly prefer an interactive experience. For those with mild- to moderate physical and/or cognitive declines, but who have high socialization needs, virtual gaming in a group setting can help to engage Socializers. It is also important to ensure that in any setting (institutional or otherwise) Socializers have access to the devices and services they need to stay in touch with their large family and friend networks so integral to their identity. Often the ICTs provided in such settings are too old and too underpowered to allow effective access to social media, or social media websites are blocked. Such policies prevent many Socializers from using ICTs to fulfill their socialization needs (more on these issues is presented in the following section).

Traditionalists have the lowest ICT skills of any of the user types. They do not go online often (if at all) and are unwilling or unable to conduct online research or fill out online forms. For Traditionalists, it is critical to identify if they have a family member or friend that undertakes these tasks for them and to determine if their digital and information needs are being met. While many Traditionalists have such individuals to rely on, some do not. Telling an older adult Traditionalist to simply "go online" may be an insurmountable and unwelcome task. If a person does not have a proxy that can access the internet, this is a role that perhaps a companion or personnel at your organization can undertake.

Guardians tend to be concerned with internet security and privacy. They are overall cautious of providing their information online. For Guardians, it is important to have privacy procedures in place and have these easily and frequently explained. In institutional care settings, Guardians deeply value ICT-free spaces where they can have conversations and deepen relationships face-to-face without distraction from digital devices.

The ICT User Typology provides insight into how we can better support older adults to improve their life satisfaction. While this section has briefly introduced how practitioners can best facilitate each type's desired use patterns, the next section examines the barriers that prevent the desired levels of ICT use. In order to discuss how these barriers can be overcome for the five user types these challenges are explored through the most common language used to conceptualize them: the digital divide.

Closing the Gray Digital Divide: Supporting Older Adult Users *and* Non-users

Oftentimes in the literature, there is a push to encourage older adults to adopt ICTs. Such literature seeks to close "gray digital divide," – the idea that older individuals report lower usage rates of the internet, computers, and cell phones compared to younger people.[2] Even younger elders show higher rates of ICT use than older elders (Heart & Kalderon, 2013). However, such an oversimplification of the gray digital divide misses the complexity in older adult ICT use and non-use. Individuals move in and out of various user and non-user categories due to changes in their physical, mental, financial, and motivational states(s): the digital divide is not static (Van Dijk, 2005). Indeed, three groups of people are being captured in the digital divide literature at any given moment: those that are using digital ICTs, those that want to (but cannot use them at the level that they wish), and those that do not want to use ICTs (Millward, 2003; Russell, 1998; Van Dijk, 2005).

The first two groups, those older adults that are using digital devices and those that want to, cut across four of the user types: Enthusiasts, Practicalists, Socializers, and Guardians. Within these four user types, you will find many individuals using the devices they would like, in the ways they would like, for the purposes they would like. However, you will also find individuals who are not able to access the ICTs they desire due to a lack of technological literacy, functional or cognitive declines, or financial barriers.

Improving Technological Literacy: Helping the Wants

Lack of knowledge, and particularly support for learning new ICTs or new functions, is a particular problem for many older adults (Morrison, 2015). Among the Lucky Few, there is a great diversity in exposure to ICTs through the work environment, so retired older adults in this generation are left with very differential ICT knowledge in elder age. For instance, both George (Guardian) and Dan (Practicalist) found that, because of their relatively high positions in organizations (a vice president and a director, respectively), they had not been exposed to

[2]The literature on the digital divide and older adults spans several decades (Czaja & Lee, 2007; Czaja & Sharit, 1998; Friemel, 2016; Gilleard & Higgs, 2008; Jacobson, Lin, & McEwen, 2017; Millward, 2003; Paul & Stegbauer, 2005; Peral-Peral et al., 2015; Pick et al., 2015; Van Volkom et al., 2014).

computers and the internet through their work. Both wanted to learn computers once they reached retirement, but found learning without a supportive work environment difficult.

Dan was starting his own consulting business following his retirement from his career in a global non-profit. One of the factors in the timing of his retirement had been the cognitive decline of his mother due to dementia. He returned to the rural area in which he grew up to serve as her full-time caregiver. Even though Dan was happy to return the loving care he received from his mother as a child, the experience was often isolating and heartbreaking. (Dan's mother died shortly after the conclusion of our interviews.) He sought to create a consulting business and to advise graduate students not only to engage his brain, but as a welcome "escape" from caregiving activities. However, his connections to these individuals were virtual by necessity and he found himself struggling to use the computer technology that he had not learned in his prior work.

Due to Dan's rural environment, there was no formal place within easy driving distance he could turn to for help. Instead he relied on his wife to help him learn. However, he found it frustrating that while he could give a fantastic presentation, he struggled to create digital slides. (See Chapter 3 for a more in-depth discussion of Dan's experience.)

For George, he found using a computer challenging. He relied on the two of his children who lived locally to provide technical assistance, although he often wished that there was someone he could hire to provide technical lessons on specific software and applications he wanted to use. While George worked part-time at a big box retailer (for, as he put it, "something to do"), the vast majority of his position did not require nor provide the opportunity to use technology. As a Guardian, there was some technology he was interested in (mostly the computer and internet) but other technologies he was not. He indicated that he would like tailored lessons, particularly on online shopping and internet security, which were more complex topics than the local library offered in their senior computing course.

Lack of training is often cited in the research as being one of the greatest barriers to older adults learning ICTs (Friemel, 2016; Padilla-Góngora et al., 2017) and research has focused on developing specific strategies to accommodate older adult learning needs (Boechler, Foth, & Watchom, 2007; Slegers, van Boxtel, & Jolles, 2007). The need for better and more customized technical training for older adults is clear. Rather than seeing the offering of more content as a challenge for already underfunded community programs, such needs can be instead seen as an opportunity to promote intergenerational understanding and communication.

As our populations age worldwide (Kinsella & Velkoff, 2001), there will be many aging-related employment opportunities (Rosowsky, 2005). These positions will not only be in healthcare but across all industries – from housing to tourism, marketing to design, and government to business (Plawecki & Plawecki, 2015). Such a population shift will require those entering the workforce to be prepared to work with an aging clientele and meet the demands of an aging market. A potential solution to meeting the demands of an aging-savvy workforce *and* to provide technical literacy to our aging population is to have young people (secondary and post-secondary students) help older adults to learn new technologies (under the guidance of teachers and professors). Such older adult technological literacy programs could be modeled on STEM programs[3] – insofar as they encourage younger people to consider a new career path – in this case an aging career path. Such experience with older adult clients and/or consumers will be considered an asset in any young person's chosen career given societal aging trends. Taking into the account the growth of older adult markets (particularly, as Boomers begin to age into elderhood (Coleman, Hladikova, & Savelyeva, 2006)), such exposure not only benefits the older adults, but also organizations seeking workers to serve an aging clientele and young people seeking employment.

When creating such a technological literacy program, it is important to involve older adults, who best understand their needs and desires (Lenstra, 2017). While many libraries and senior centers have offered computer classes for older adults (Anger, 2005; Eaton & Salari, 2005; Xie & Jaeger, 2008), one of the issues of this common model is the lack of customizability. Practicalists and Guardians are not interested in learning every possible function they can complete with their computer and would prefer learning about targeted features. Dan particularly wanted to learn office software as he was interested in starting his own business. George wanted to learn about information security and online shopping.

Research has shown that private, one-on-one tailored technical training is highly desired by older adults (Friemel, 2016), so such lessons should focus on instructor and learner (rather than group) environments. Since both emotional and technical support are important to older adults successfully learning new technologies (Barnard, Bradley, Hodgson, & Lloyd, 2013), it will be important to train volunteers to recognize and be sensitive to older adults' feelings toward technology.

While addressing technological literacy is important to facilitate many of the user types, there must not only be attention paid to providing knowledge, but also accessible *devices* that allow individuals with a diversity of ability level access to our digital society.

Facilitating Accessibility to and Affordability of ICTs: Helping the Wants

ICTs are typically created for the young (Larsen, 1993): people with good eyesight, high manual dexterity, and fast cognitive reflexes and memory (Becker, 2004). Many older adults experience cognitive (Fang et al., 2017) and/or physical declines or disabilities which make using modern ICTs difficult (Barratt,

---

[3]See Breiner, Johnson, Harkness, and Koehler (2012) for a discussion of STEM programs and partnerships.

2007; Czaja, Sharit, Charness, Fisk, & Rogers, 2001; Hill, Betts, & Gardner, 2015). While in many high-income developed nations the chronological time older adults are spending disabled at the end of life has decreased, functional limitations still begin to rise in the older population at age 70. For low- and middle-income countries, such limitations increase at even younger ages (Chatterji, Byles, Cutler, Seeman, & Verdes, 2015). Functional limitations are, and will continue to be, a major stumbling block impacting older adults' technology use in generations to come.

Across the study, individuals expressed concern that if they developed such limitations, or if their current impairments progressed, they would find it difficult to use the devices they currently used. Such concerns are not imaginary: older adults who face physical or cognitive impairments report much lower usage rates of ICTs than those without such impairments, with vision and memory declines being most impactful (Gell, Rosenberg, Demiris, LaCroix, & Patel, 2015). Those who have such impairments are often fearful that new innovations will exclude them (Okonji, Lhussier, Bailey, & Cattan, 2015).

Older adults need accessible devices that are sensitive to their lifestyle, their needs and wants in technology (Bagnall, Onditi, Rouncefield, & Sommerville, 2006). The ICT User Typology can help us to identify the wants and needs in the diverse older adult population and we can tailor and build devices to meet their lifestyle preferences – while ensuring such devices and services are accessible. The physical and cognitive inaccessibility of technology can prevent Enthusiasts, Practicalists, Socializers, and Guardians from using the technologies they desire.

While one of Gerontechnology's main goals has been the creation of alternatives for those with disabilities or impairments to use modern ICTs (Bouma, 2001), attention must be paid to ensure these devices are financially in reach. Nancy is a prime example of how inaccessible technology design can impact an older adult and prevent them from living their ideal lifestyle. As a Socializer, she wanted to be able to text, to connect her to the youngest individuals in her large intergenerational network. Living in an assisted living facility, she had no disposable income and no resources with which to purchase an alternative device and/or service that would allow her the functionality of texting. While Nancy qualified for the US government Lifeline program (which provides qualifying individuals with a free cell phone and limited monthly free minutes and data) (Federal Communications Commission, 2018), all of the available simple phone models had small buttons. The last phone she had tried she attempted manipulating with a pencil, however, even the eraser was too large. She sent this phone back, and refused to participate in the program again, as she said it was not designed for people like her – people with impairments or who were older.

One of Nancy's fellow residents, Bobbie, showed me her Lifeline phone model. A simple phone (non-smartphone) was smaller than a deck of cards (Figure 16). Although Bobbie was significantly younger than Nancy and had no eyesight or manual dexterity issues, she found the phone difficult to manipulate. In Figure 16, the cell phone is the smaller device. For size comparison, you can see the size of a normal television remote and several hard candies in a plastic

Figure 16. An Example of a Lifeline Phone.

sandwich bag. I handled the phone and attempted to dial my own number – and it took two tries to enter it correctly because of miss-struck keys.

Many older adults receive federal social welfare benefits in the US and, because of this, they qualify for a Lifeline phone. It is disheartening that the only phone models commonly available are often unsuitable for older individuals and/or those with disabilities. For Nancy, who mentioned many times over the course of the study that she just "wished they had a phone I could use to text," the frustrations of not having an accessible device were heartbreaking. As she pointed out – it is not only older people who have arthritis or difficulty seeing who needed to use a cell phone. Such devices, too small to manipulate, disenfranchise large swathes of our societies.

While individuals with more disposable income may be able to compensate for their disabilities or declines through self-purchasing, it is concerning that the devices provided for our most vulnerable society's members are often unusable. When researchers and practitioners speak of closing the gray digital divide, it is important that we remember to not only make ICTs available for those who do not have access – but also to make them physically and cognitively accessible for an aging population.[4] More resources must be invested in designing more age-friendly ICTs. Importantly, there should be a focus on making sure that the cost of these devices is kept low and they are available on appropriate governmental assistance programs.

---

[4]Many researchers have outlined methodologies for designing with older adults with cognitive and/or physical impairments (Astell et al., 2009; Bagnall et al., 2006; Dickinson & Dewsbury, 2006; Dickinson et al., 2004).

Prior research has indicated that inability to afford ICTs impacts older adults' self-fulfillment (Hill, Beynon-Davies, & Williams, 2008). It is no wonder: technology for Enthusiasts is the center of their lives and for Socializers is how they stay connected to their important communities. Denying these user types devices because they have developed disabilities or age-related impairments is to take a strike at their fundamental identity. Studies have shown that communication technology use decreases loneliness (Czaja et al., 2018) and social isolation (Blit-Cohen & Litwin, 2004; Chen & Schulz, 2016; Clark, 2001; Czaja et al., 2018; Xie, 2008), while improving well-being (Blit-Cohen & Litwin, 2005; Czaja et al., 2018; Ihm & Hsieh, 2015; Khvorostianov, Elias, & Nimrod, 2011). Some research has suggested that among retired older adults, internet use is even correlated with lower depression rates (Cotten, Ford, Ford, & Hale, 2012). Designing, creating, and providing accessible and affordable technologies for those who put technology or communication at the center of their lives should be paramount.

Even if devices are accessible and affordable, they need to be available to older adults in their living situations. In our resident care settings, policies and dated technology can prevent older adults from engaging with our digital world. Gerontechnologists' work in such settings has often looked toward providing care or workforce solutions (Czaja, 2016; Freedman, Calkins, & Haitsma, 2005), rather than in facilitating residents everyday use. In Nancy's assisted living center, access to social media sites was blocked in order to prevent staff members from wasting organizational time. This meant that residents that wanted to use social media had to travel elsewhere to do so. Most residents were unable to travel to the nearest location that offered both free Wi-Fi and computers: the local library. A mile away and not located on any public transportation line that connected to the assisted living center, this was inaccessible to the vast majority of residents.

Providing access in such environments makes sense as growing numbers of older adults rely on social media as a way to remain connected to their families (Coelho & Duarte, 2016) and access to communication technology has been proven to ease the transition to such living communities (Waldron et al., 2005). Many of the residents of the facility I spoke with would have enjoyed access to social media; however, they were prevented from doing so because it improved overall organizational efficiency. There are a multitude of other interventions that could have been used to prevent employees from using social media during their workday: stricter policies, blocking individual devices, and collecting devices from employees during their shifts. Policies which facilitate residents' digital lives and help them to more closely achieve their ideal technological lifestyle should be a priority. The residents, as clients, should be the center of care.

Beyond allowing access, available technology in these facilities is often sluggish and outdated (or non-existent). The computers available in Nancy's assisted living facility were over 10 years old. It should be an institutional priority to provide a number of relatively recent computers to residents. These computers need not be purchased new. Individuals and organizations frequently update devices, with organizations in particular often only receiving small recycling returns.

Such devices, rather than being returned to the manufacturer could be donated to such facilities with a charitable deduction. Many organizations with older adult clients may find student groups at their local secondary school or university willing to help source devices and provide technical help.

While increasing accessibility and access are important to allowing Enthusiasts, Practicalists, Socializers, and Guardians to use ICTs in the ways they desire, it is also important to consider how to best facilitate the Traditionalists – the want-nots – in our society as well.

Supporting the Traditionalists: Helping the Want-Nots

The third important category of older individuals in the digital divide is the "want nots" – these are people choosing to not use modern digital ICTs (Eynon & Helsper, 2010; Russell, 1998). In terms of the ICT User Typology, these are the Traditionalists – those individuals who love the ICTs of their youth but have little interest in using more modern technologies.

Much of the focus in both popular media and in scholarly and practitioner digital divide literature focuses on how to motivate Traditionalists to use the modern ICTs they reject. The presumption in much of this literature is that use is good – and non-use, conversely – is, therefore, bad (Friemel, 2016; Russell, 1998). In these discussions of the digital divide, we often overlook older adult personhood (Lenstra, 2017) seeing technology as the perfect "solution" to the "problems" of aging (Parviainen & Pirhonen, 2017). Such insinuations suggest that aging is not valued, nor desirable, but instead is a state that needs to be fixed. Our aging societies do not need fixing, technology is not a bandage, nor is any technological innovation free of consequences – be they positive or negative.

Older adults have a right to self-determine their level of engagement in our digital world, including if that level of engagement is zero. They have the right to make choices about their lives and this includes the choice to disengage from more modern forms of technology. The focus on encouraging computer and internet use in much of the literature infantilizes older adults, playing into ageist stereotypes which suggest that they are not capable of making their own decisions (Cutler, 2005). Older adults are not children and encouraging them to use a computer should not be analogous to telling a toddler that they should eat their spinach.[5]

Older adults are well aware of what technologies can do. Traditionalists, who consume high amounts of traditional media, are well aware of many new

---

[5]In the United States, there is much fascination with religious subcultures that have rejected and/or restricted their ICT use, including the Amish. Our acceptance, and even fascination with individuals from a subculture that rejects and/or strictly controls their ICT use (see Umble (1994) for a discussion of the telephone), stands in contrast to our societal non-acceptance of the older adults among us who choose to not use ICTs.

technologies: they have seen/read/listened to advertisements, stories, articles, and programs about them and still they have no interest in using these modern forms. Traditionalists have little self-motivation to use such technologies in their elderhood, as their lives are simply full of the traditional media they love. In many cases, Traditionalists are already being pressured by friends and family members to adopt more modern ICTs. They find themselves inundated with well-meaning but unwanted technological gifts, yet they continue to resist this pressure to use them. With Traditionalists' more modern technologies gathering dust and frequent inquiries from loved ones, there is not much more on a societal level that can be done to encourage Traditionalists' use.

Those of us who identify as Enthusiasts cannot let our love and passion for technology blind us to the agency of our elders. If we would like to have agency in the final decade or two of our own lives and to be able to make our own choices about the technologies we use, we should start by respecting the technological choices of our own elders now.

Traditionalists from the Lucky Few generation are mostly unharmed by their decision to disengage from the virtual world. They tend to be very happy about their choices (although they are often frustrated by the pressure they feel from others to use modern technologies). These individuals have survived, adapted, and thrived for over 70 years; if they are happy being unengaged from modern ICTs in the last decades of their life, there is a strong argument that we should allow them to do so.

There is an important caveat to this discussion. Much of our information and services are moving online, leaving people who are not internet users at risk. Successful older adult Traditionalists tend to have a strong network of people who use ICTs and they rely on these direct users to get them information they may need, file things online, or digitally communicate with others. These direct users are similar to the "warm experts" previously identified in the literature – individuals who provide technical assistance and/or access to older adults (Wyatt, Henwood, Hart, & Smith, 2005). In this way, Traditionalists are "indirect" or secondary users of online services, computers, and other digital technologies, relying on these warm experts for actual use.

If Traditionalists lose their relationship with the direct user they rely on, they are at extreme risk in a digital society. Mindy Jean relied on her husband to go online, including to file taxes and look up information. June relied on her children and a man who worked at the front desk at her low-income apartments to look up information and print off forms she needed. If these Traditionalist older adults were to lose these individuals, or never had these contacts in the first place, they would be at risk of not being able to complete these tasks. This speaks to the need for our societies to provide services that facilitate "indirect" users of systems: a social safety net for the indirect users of the internet. When asked what they would do if they no longer could rely on their direct users, Traditionalists volunteered they could go to their local library and the librarian would help them to get the information or to sign up for the services they needed. These types of library services need to be funded.

To truly facilitate the needs of Traditionalists, professionals and volunteers in such programs need to be willing to set aside any "pro-technology agendas" (Margaret) and ageist stereotypes. Often, particularly among those who are Enthusiasts, there is a tendency to act as technological evangelists, to encourage digital technology use and to view such use as the desirable end goal. Such a pro-technology agenda is unnecessarily harmful to Traditionalists. Traditionalists often feel shamed for their choices in ICT use, and any program that hopes to engage older adults in our digital society should avoid passing any such judgment.

In a similar vein, it is also important to support Guardians' choices in using ICTs in a limited way. While Guardians are far more willing to use more modern ICTs than Traditionalists, they tend to be very cautious in their use. Due to their knowledge of potential technological pitfalls when it comes to information security and privacy, many simply need reassurance that the ICTs they want to use are safe. Such reassurance must occur in verbiage that is appropriate for the individual Guardian's skill level. Technological language, particularly when used to speak down to Guardians, tends to be very off-putting and causes disengagement. However, there are numerous examples of Guardians being "coached" out of their original comfort zone by kind and understanding IT professionals. Margaret began online shopping after a positive experience with an employee she contacted on the phone, who described in detail (in terms she could easily understand) the security in place for online purchases. While Margaret was certain she would buy very few items online, she purchased exclusively from this online retailer as she understood their security procedures. As a result of a retail employee taking the time to discuss Margaret's concerns about privacy and information security, the company won a life-long dedicated customer.

As a society, we need to accept that choosing to *not* use a technology is a valid response. A person's value and their ability to engage in society should not be based on their technical expertise or their willingness to use the latest gadget. It is not Traditionalists and Guardians that necessarily need to change their views, but perhaps rather the rest of us.

Addressing concerns over ICT literacy, usability, and access are important to facilitate older adults using ICTs in ways that come naturally to their user type. There is also an opportunity for those who are interested in appealing to the graying market to use the ICT User Typology to better understand, design for, and market to the older adult population.

Using the ICT User Typology for Tailored ICT Service and Product Design

As our population ages, the older adult market for technological devices and services grows. Gerontechnological researchers are just beginning to recognize this potential (Hough & Kobylanski, 2009; Kashchuk & Ivankina, 2015), although there have been very few studies from the business or marketing perspective

(Mostaghel, 2016). Technologies for an aging population not only need to be adaptable to overcome cognitive and physical limitations and disabilities, but also must keep in mind the diverse motivations, needs, desires, and lifestyles of older adults (Bagnall et al., 2006; W. A. Rogers & Mitzner, 2017). As a diverse group there is no single "older adult" design that will appeal. Instead, such design must be tailored to this diversity (Righi, Sayago, & Blat, 2017). The ICT User Typology segments the older adult market, allowing targeted design and advertising to these differing tastes, wants, and lifestyles; unlocking the meanings older adults apply to technologies. It is meanings that are critical to understanding older adult technological acceptance (Hauk, Hüffmeier, & Krumma, 2018).

Each of the user types finds different aspects of ICTs interesting to them, and as a result, have different design specifications that appeal. The ICT User Typology provides guidance in designing and marketing effectively to the older population as a diverse group of individuals, rather than ineffectively to older adults as a single market. Table 3 provides guidance on what appeals to each of the user types in devices and services, how these products can be best framed for marketing purposes, the level of support that will be necessary to engage this part of the older adult market, as well as main takeaways.

Enthusiasts love technology and, in particular, love its fun aspects: they love to experiment with a new "toy," and they are very much self-supported. To appeal to Enthusiasts, it is important to make sure that any product, device, or service you offer truly is fun to use. Enthusiasts are geeks (and they are proud of it!). They enjoy showing off this geekiness. It is very important when looking to appeal to Enthusiasts that you do not talk down to their skill level. Many Enthusiasts are or were IT professionals, and many of them built technologies and networks. They do not need to be told how to use a device, but rather they are very capable of being able to open a box or subscribe to a service without much help at all. Because technology is central to their lives, Enthusiasts with impairments or disabilities represent a large market for adaptive technologies that facilitate their continuance in using digital devices. They are the most likely of the types to adopt smart home or ubiquitous computing and embrace technological solutions for age-related declines.

Enthusiasts are researchers and base their purchases off their own research as well as feedback and recommendations from fellow Enthusiasts. They pay attention to advice from their technical friends and read various technical blogs and magazines. They have the skill set necessary to use popular devices and are uninterested in technologies that will stereotype them as "non-technical people." Since Enthusiasts place ICTs in the center of their lives and throughout their homes, functional beauty is very important to them. They want their technologies to look sleek and beautiful; to be showpieces. They love their ICTs and the fun they have using them, but they do not like ugly things in their living spaces.

You may be wondering why it is important to market to Enthusiasts, who are likely to try most technologies on their own. Such marketing is critical as Enthusiasts spend a disproportionate amount of their income on technology and technological services compared to the other types. Appealing to their tastes is important. Technology is, after all, their main hobby (and often their job as

Table 3. Designing Products and Services for the Five User Types.

User Type	Is Drawn to	To Sell to	To Support	Takeaways
Enthusiasts	Fun, play, newness, experimentation	Emphasize the fun Present use as play	Requires little support; can "turn on and go"	Enthusiasts are not trying to be hip, they genuinely want to use the latest and greatest
Practicalists	Functionality, practicality, usefulness	Emphasize the functionality Present use as practical	Requires substantial support; prefers documentation and help services	Practicalists are not explorers; you must prove to them it is useful and how it is useful to them
Socializers	Connection, community, relationships, socialization, Engagement	Emphasize the connectivity Present use as bridging	Requires little support; relies on large social network	Socializers want what young people have; facilitate their ability to use what young are using
Traditionalists	Nostalgia, technology and media of their youth	Emphasize the nostalgia Present use as comforting and traditional	Needs technology that functions like old standbys	Traditionalists love the older ICT forms; still a market for new devices that "function" like the old
Guardians	Security, control, relationships, unobtrusiveness	Emphasize the security of device/service Allow individuals to control their own use Unobtrusive devices	Requires substantial support; reassurance technology is safe	Guardians are concerned that technology is all-consuming and unsafe; give them the ability to control their use

well). Enthusiasts are well versed in the technological landscape and are well aware of your service and product – and your competitors. They are often the "technical help people" among their family and friends, no matter their age. When Enthusiasts recommend a device, the other user types listen. Earning older Enthusiast customers, therefore, can help to bring in other user types as customers as well.

Enthusiasts greatly dislike advertising that presents older adults as non-technologically capable. Specifically, they dislike advertisements that suggest that older users have no technical skills or need simplified technologies. These they view as very stereotypical and are unlikely to try any technology they feel is oversimplified and dodgy. Older adult Enthusiasts, however, are not just carbon copies of younger people with the same tastes and preferences. When asked, many Enthusiasts would like to see more "positive" portrayals of older adults in technological advertising – portrayals that match their skill levels. Older adults who are portrayed as capable technology users having fun using the device, service, or product will particularly appeal to older Enthusiasts.

Practicalists are focused on the functional and practical aspects of the ICTs they use or are considering using. They do not play with ICTs, they do not experiment, and they do not see technology as fun or a toy. Technology is a tool. Appealing to Practicalists involves outlining the functional aspects of a technology – how it can be used in Practicalists' daily lives and how it is an improvement over their current technology. Practicalists are drawn to devices that offer greater functionality and usefulness than their current forms. They tend to be very definite in the ways they want to use a technology and have little interest in devices that stretch across every aspect of their lives. Practicalists' tend to see certain ICTs as leisure devices, and others as work tools or community builders.

Practicalists strongly dislike (even to the point of "hating") having to search out new features or uses; they do not play with their devices – their devices are tools. It is important, therefore, that if your device or service has features which are not readily apparent but are highly functional that you use walkthroughs to show Practicalists how to use these functions on their newest tool. Practicalists do not experiment to discover "cool things" your ICT can do; they are typically told, so be sure to do the telling through various support documentation. A lack of such documentation frustrates Practicalists. They will not open a device or service and begin poking around; they need and desire some level of guidance. If a competitor proves that their device is more easily used and more useful, you have likely lost Practicalist customers.

Practicalists need a high amount of support, particularly since they do not experiment with ICTs. When they run into issues they need a place to turn to for answering their questions. This can include support such as a help line, manuals, and/or web support. Practicalists often have relatively high skill levels in those technologies they use frequently, so the need for such support is not necessarily remedial. Instead, if Practicalists have questions about a device or service, such as how to change a setting that is not readily apparent, they would like to be able to look such information up, rather than exploring how to do it on their own by selecting various menus or icons and experimenting. Practicalists are frustrated when they receive an ICT with no manual or instructions on how to find help online. Technologies should be easy to use and help should be readily available in Practicalists' minds.

Many Practicalists were exposed to ICTs in their work. They often adopt versions of these ICTs for their home life, if they see a relevant application. They are not impressed by others simply owning a device, or a device being "hip" or "popular." They want to know a device works, and for what purposes, and seek out advice from friends and coworkers, as well as information in blogs, news articles, and reviews that help them determine the functionality of a device. They are uninterested in advertisements that are unsubstantial and do not detail functionality.

Socializers are focused on the connectivity of their devices. They love to socialize, meet people, and build connections with others. Of all the types, they are most likely to be influenced in their use by those around them. They have large intergenerational networks and, as a result, tend to adopt the ICTs being used by the young people in them. Many older adult Socializers text (or would like to text), but those who have arthritis or other disabilities (such as visual impairments) are prevented from doing so. Creating adaptive devices that allow Socializers with impairments to mimic the way young people communicate would do much toward facilitating their everyday lives and be highly appealing to this group. Socializers' interests are highly geared toward technologies such as social media and mobile communication technologies such as smartphones, tablets, and other portable communication devices.

Socializers are the most likely to find advertising targeted toward younger individuals appealing. As a result of the intergenerational nature of Socializers' networks, they are always observing which technologies are being used by and marketed toward their youngest family members and friends. They place little to no value on technologies, devices, and services that isolate; greatly preferring those that enable connection. Much of this value is not inherent in the ICT itself, but rather in how the use of the ICT is framed. For instance, while video games and other forms of digital gaming are largely seen as an isolating activity, those which can be played in a group setting and are marketed as a social activity can be quite appealing to Socializers.

Socializers do not use the devices young people do to be young or hip. Instead, they understand that they are adopting the ICT use patterns that have been established by younger people. Learning these patterns is extremely important to Socializers, as they credit these technologies with strengthening their relationships. Marketing that focuses on creating and building bonds with their large families (such as with their grandchildren) is particularly effective. They have no interest in nostalgia or stigmatizing devices created solely for older adults.[6]

---

[6]Stigmatizing adaptive technologies are unwelcome among older adults on the whole (Gitlin, 1995; Yusif & Hafeez-Baig, 2016).

Traditionalists are the most likely of the five user types to focus on nostalgia. They fill their lives with the more "traditional" forms of media they encountered when they were young. Traditionalists are the most difficult segment of the older adult population to market new devices and services directly to, given their resistance toward using new technologies.

There are two opportunities for those who are interested in tapping the Traditionalist market, however. First, highly nostalgic, Traditionalists deeply enjoy using the media of their youth and young adulthood. This includes the music, television, and movies from their younger days. There are a multitude of opportunities to market such nostalgic experiences to Traditionalists, particularly if such experiences are coded within technologies that have the appearance and user interfaces of older ICT forms. While older adult Traditionalists would not be willing to adopt a digital music player presented as such, they would be keen to purchase a player that looked and acted like a regular radio but was pre-loaded with their choice of nostalgic music. Presenting nostalgic technologies that most importantly *act and function* like the ICTs Traditionalists know and love is incredibly important to appealing to this user type. These nostalgic devices can also be marketed to the family and friends of Traditionalists, who are often keen to introduce the Traditionalists in their lives to more modern ICTs, but find their efforts rebuffed. Devices which mimic the functionality of the Traditionalists' beloved ICTs of their youth while incorporating new technologies may be a suitable middle ground. Second, while Traditionalists are typically not primary users of modern ICTs (when at all possible), they often rely on others to access online information and services. There is an opportunity to provide these services to Traditionalists who do not have a friend or family member to access this information directly.

Guardians are interested in ICTs that are secure, can be easily hidden, and are controllable. Unlike Enthusiasts, who are in love with ICTs not only for their function but form, Guardians enjoy having ICT-free spaces. They have absolutely no interest in ubiquitous computing, and the idea of smart homes that they do not directly control is frightening to them. They like to have absolute control of the technology they use – including the ability to turn it off and place it out of sight. For designers, there are opportunities to develop ICTs that are easily hidden that retract into spaces or mimic other objects.

Guardians have a large diversity in skills, ranging from beginners to more advanced users. Since they tend to be naturally hesitant to try new technologies and place emphasis on face-to-face interaction, they tend to need high levels of technical help and support. In particular, they prefer having as contact-rich methods of seeking help as possible – being able to speak to someone over the phone or in-person is preferred to a web-based chat.

When marketing a device or service to Guardians, it is very important to ensure that the technology is secure[7] and controllable. Guardians need

[7]Overall, for older adults, security is very important for their use of ICTs, and these security protections must be clear and easily understood (Jung, Walden, Johsnon, & Sundar, 2017).

explanations of how their information will be protected and reassurance that the technologies are secure and private. Any such information should be an important part of any marketing plan, as they avoid ICTs which they believe risk their privacy or security. Similarly, they resist investing in smart appliances, as they are well aware of the potential information security threats these devices pose.[8] In developing ICTs for Guardians, it is important to develop secure technologies. Once a Guardian has had their security breeched, they will be unwilling to continue using your device or service.

While the ICT User Typology informs more targeted design and advertising to the older adult market, it also has practical implications for those interacting with younger individuals, due to user types developing early in the life course.

Moving Beyond Older Adults: Using the ICT Typology in Primary and Secondary Education

The ICT User Typology highlights the importance of child and young adulthood experiences to the development of a person's user type. Enthusiasts credit strong positive childhood memories of ICT use, being encouraged to tinker, and having a technological mentor as being critical to developing their love of technology. An open question, however, is if such childhood experiences can be manipulated to influence a person's user type. If so, such experiences can be designed to encourage larger portions of Enthusiasts in the population, but also possibly inoculate against other user types developing.

There is a well-documented need for STEM (Science, Technology, Engineering, and Math) professionals (Directorate-General for Internal Policies, 2015; Gonzalez & Kuenzi, 2012; Kuenzi, 2008; Marrero, Gunning, & Germain-Williams, 2014; Microsoft, 2017). Creating tinkering environments with technological mentors for young people could influence user type development, resulting in more Enthusiasts in the population.[9] Such exposure to the many potential uses of ICTs could also result in a higher percentage of Practicalists, given the importance of understanding application to this type. Since both Enthusiasts and Practicalists are likely to be involved in careers which use a large amount of technology, there are many potential advantages to increasing their numbers. Indeed, evidence exists that providing early positive technological experiences, messages, and mentorship can prevent or inoculate individuals from developing negative technological viewpoints (Stout, Nilanjana, Hunsinger, & McManus, 2011) and, therefore, such programs could prevent many Guardians from developing their cautious attitudes toward ICTs in later life.

---

[8]See Rowe and Trejos (2017) for a discussion of smart appliances and information security risks.
[9]Evidence already has suggested the importance of mentors in influencing interest in STEM (Tillinghast et al., 2017).

Many programs are already in place that are trying to accomplish the goals of providing tinkering opportunities and technological mentorship to children and young adults. These include primary and secondary programs (Directorate-General for Internal Policies, 2015; Gonzalez & Kuenzi, 2012; Kuenzi, 2008) and community-based programs (Tillinghast et al., 2017), which are often aligned with higher education institutions (Breiner et al., 2012). Such programs should be encouraged. It would be of immense value to institutionalize these programs in primary and secondary schools, free-of-charge to participants.

However, we must recognize that each of these user types has a place and role in our society. Enthusiasts are our eager innovators and Practicalists our productivity-focused colleagues. Socializers connect us generationally. Traditionalists provide us with a sense of a grounded past while Guardians are protective watchers of our society and safety. Life without all of these five types would be boring and unbalanced. We need protective Guardians who are watchful of privacy concerns and rights online as much as we need innovative Enthusiasts; we need Traditionalists in order to be grounded in our historical context as much as we need Practicalists' functional focus. And, as our societies become more age diverse, it is more than nice to have a few Socializers who keep us all connected and communicating intergenerationally.

While it is unlikely that technological education and mentorship would cause the extinction of some of these user types, the goal of such education should not be to extinguish any one type through social engineering. Such programs should, instead, be designed to meet the needs of an ever-increasingly digital workforce, while recognizing the value these diverse user types bring to our societal tapestry. Variety in user types, as in all things, is the spice of life.

The final chapter of this book provides a detailed discussion of the methods used to generate the ICT User Typology and provides helpful hints to researchers seeking to replicate this study in other cultures or populations. It also discusses some of the lessons learned working with older adult participants and provides a detailed discussion of how to use the interpretative interactionist methodology to develop theory and theoretical constructs.

Chapter 11

Discovery of the ICT User Typology

During a research project over a decade ago, I became inspired by the older adults I met in a community center computer class.[1] These diverse older adults had come to learn new skills, but often had to overcome the challenges of being a new learner, sometimes combined with physical and/or cognitive declines or disabilities. Through this class I met Margie, with whom I recorded a life history. Margie and I spent most of a Friday together while we spoke about the influence of technology on her life. She, along with her husband, had been involved in organizing labor in the city where she had lived during the Great Depression. Both she and her husband had risked life and limb to do so: her husband had been beaten and they both had been jailed numerous times. Beyond her fascinating life experience, however, she also taught me about what it was like to navigate an often youth-obsessed culture filled with a bias toward older adults.

Margie explained to me that although a person "could find out a lot of information about what it is like to be old, most of that information comes from people who aren't old." Such refrains, while spoken about (Bragg, 2004), have not often been reflected in the ways we collect data in the empirical gerontechnological literature (Birkland & Kaarst-Brown, 2010). Gerontechnology has often approached studying ICT use by focusing on external measures of older adult performance or simple adoption studies, not on how older adults feel about using technology: the meanings they ascribe to the ICTs, their use, and their rejection (Brophy, Blackler, & Popovic, 2015). While these prior studies have great value; they often miss the older adult's voice. The feeling of being voiceless that Margie expressed would be echoed by many older adults I have met over my career.

After meeting Margie, I saw my mission in my research changing. It was not only about sharing the experiences of older adults regarding technology, but also about sharing their voices. For my work, showcasing older adults' experience through their own words became paramount: I wanted to tell stories told by the older adults who lived them.

The following sections denote the method I used to discover the ICT User Typology. Provided both for replication and greater understanding of how the typology was developed, it demonstrates how the evidence supports the claims made in this book. Beyond this evidentiary purpose, it is intended to guide those

[1]In this chapter, I refer to myself strongly in the first person to carefully delineate my agency and choices in methodology.

interested in interactionist methodology and/or those studying older adults. I've chosen to present this chapter as a journey, which allows you, as the reader, to see the messiness of the process, the issues I encountered, and the active choices I made as a researcher.

The Process of Discovery: Determining the Method

Early in my search for a plausible method to explore older adult ICT use, I investigated many potential methodologies, mostly qualitative. The original goals of the study were not to create a theory, but rather an explanation. I intended to document the diversity of older adult ICT use and begin to carve away at the reasons why such diversity existed.

In organizing my thinking, I explored many common theories of ICT adoption, including the Technology Acceptance Model (TAM) (Davis, Bagozzi, & Warshaw, 1989), UTAUT (Venkatesh, Morris, Davis, & Davis, 2003), Diffusion of Innovations (E. M. Rogers, 1962, 2003; E. M. Rogers & Shoemaker, 1971), and Domestication Theory (Silverstone, 2007; Silverstone & Haddon, 1996; Silverstone & Hirsch, 1992; Silverstone & Hirsch, 1994). TAM, UTAUT, and Diffusion of Innovations have commonly been explored to understand technology adoption and use by Gerontechnologists (Charness & Boot, 2016). Based on my background research in older adults and ICTs, I found that these commonly employed theories failed to explain the diversity in use, as I was interested in use more than in adoption.

When I encountered Domestication Theory (Silverstone & Haddon, 1996; Silverstone & Hirsch, 1992), it immediately spoke to me. It conceptualized ICT use in a way that I felt was logical but, more importantly, it gave me a framework to study the introduction, use, display, and meanings of ICTs, while also examining non-use (Umble, 1994). A relatively new theory to the gerontechnological literature (De Shutter, Brown, & Abeele, 2015), it focused on qualitative data collection methods I knew would elegantly showcase older adult voices.

Those studying domestication most often employ a case methodology (Haddon, 2007) where ICT use is bounded, be at the country, social group, household, or in the case of this study, individual level. Treating each older adult as a case of ICT use allowed me to bound such use, while capturing what Flyvberg (2006, p. 223) calls the "nuanced reality" that individuals inhabit: contextual spaces that are complex, rich, and real that represent everyday life.

Early in my conceptualization of ICT use, I realized that I needed to study a diversity of ICTs to develop a more thorough explanation of use and non-use. Prior studies of a single technology did not allow us to understand the integration of ICTs used together, nor did they allow us to understand a mix of rejection and acceptance by the same person. Inspired by a description of the everyday aspects of older adult's lives determined by Gerontechnologists (Bouma, Fozard, Bouwhuis, & Taipale, 2007; van Bronswijk et al., 2002; van Bronswijk et al., 2009) I decided to examine ICT use in older adults' family, work, community, and leisure lives.

Interpretive Interactionism (Denzin, 2001) was integrated into the study as the primary methodology to understand the meanings presented in each case. A methodology which finds its roots in phenomenology focuses on finding and negotiating meanings through conversations. The flow of an interpretive interactionist interview feels much like an engaged conversation, enabling participants to feel more comfortable in sharing their views with a researcher compared to other question and answer formats occasionally seen in qualitative work (Denzin, 2001).

Having decided to explore the domestication of ICTs in older adults' everyday lives through interpretive interactionism, I turned toward determining the structure of the individual cases.

Deciding a Basic Case Structure: Multi-interview Format

Given the amount of material I was seeking to collect (the ICTs the older adult was using, not using, how they were using them, why they were using them, and their history of ICT use) a multi-interview format made practical, in addition to methodological, sense. A hallmark of interpretive interactionism, the multi-interview format, not only allows meaning to be checked in the interviews themselves, but allows the researcher time to analyze the interviews, and bring these interpretations back to the participant, resulting in further iterations of meaning-making (Denzin, 2001). I settled on a series of three interviews, as Wengraf (2001) suggests three such interviews not only result in more detailed data, but also improve participant recall.

In addition to the three interviews with the primary participants, I decided to include secondary participants (whose contribution was highlighted in Chapter 8). Secondary participants were friends, family members, and coworkers of the older adult primary participants. Secondary interviewees were incorporated for two main reasons: first, their inclusion acted as a triangulation method for establishing primary participants' ICT use, and second, I wanted to understand how technology was used in the primary participants' relationships with others.

Primary Participant's interviews lasted from an hour and a half to over three hours in length, with approximately 5–10 hours of interviews per case. Secondary participants were interviewed between the second and third primary participant interviews. Analysis after each interview was brought back to the next interview with the primary participant. If possible, 2–3 secondary participants were interviewed for each case.

Having determined the basic case structure, the next step was determining who I would interview as primary participants.

Who to Interview: Conceptualization of the Sample

Prior age-based research, which has grouped older adults into one group (often sampling everyone age 65 and older), has clouded the generational issues that can impact ICT use, including lifetime exposure to ICTs (Birkland & Kaarst-Brown, 2010). Birth cohorts, or what we commonly call generations, are groups of individuals born closely together who experience the same historical events at a similar age and life stage (Eyerman & Turner, 1998). Technology, and its introduction, is such a historical event (Birkland & Kaarst-Brown, 2010; Edmunds & Turner, 2002; Larsen, 1993). A technological introduction impacts individuals differently based on their life stage: young children have a vastly different experience than those who are older adults when the same ICT is introduced. For the Millennials (born 1983–2001), the personal computer (introduced in 1984) has always existed, while for the WWII generation (born 1909–1928), the computer was introduced during their retirement. These two generations not only had drastically different experiences with the computer, but our society views one of these generations as "natural" and "legitimate" computer users (the Millennials) while viewing others as less legitimate (the WWII generation) (Birkland & Kaarst-Brown, 2010). Selecting a single generation to study in-depth as primary participants prevented these generational differences from clouding any possible results.

In choosing a specific generation, there were several factors to balance. I was quite interested in understanding how work impacted ICT use. Work remains a critically understudied area of Gerontechnology research (van Bronswijk et al., 2002; van Bronswijk et al., 2009), with most research studies focusing on how having older adult workers impacts organizational productivity (Charness, 2006; Charness, Kelly, Bosman, & Melvin, 2001; Czaja & Sharit, 1993, 1998).[2] Such work misses how older adult's ICT use, meanings, and technical skills impact their employment.

Outside of the productivity stream of research, the presumption by many researchers appears to be that older adults are retired and no longer working. This does not always reflect reality: many older adults in the United States continue to work beyond age 65: 27% of older adults age 65–69 and 15% of older adults age 70–74 are working for pay. In fact, the rate of workforce participation only drops to 5.8% for those age 75 and over (US Bureau of Labor Statistics, 2018). Clearly, with over one in four older adults age 65–69 working and one in 20 older adults still working at age 75 and beyond, not all US older adults are retired.[3]

While many individuals in the US plan to remain working past age 65 (Benz et al., 2013) and are protected in doing so for most careers by the Age Discrimination in Employment Act (ADEA) (US Equal Employment Opportunity Commission, 1967), this is not the case globally. Some European countries have age-based mandatory retirement ages (typically age 60 or 65),

[2]Given how much of the general, non-technological focused literature on aging and work focuses on how the aging workforce will impact organizations (Burtless & Quinn, 2001; DeLong, 2004; Hedge, Borman, & Lammlein, 2006), such a focus in gerontechnological research on work is not surprising.
[3]Since the majority of middle-aged US adults plan to continue working beyond the traditional age of retirement (age 65) (Benz, Sedensky, Tompson, & Agiesta, 2013), work will continue to be an important context to study.

and such age restrictions have been upheld in court (Bilefsky, 2007). This may account for some cultural perspectives not exploring aging and work. Given that this study was to be conducted in the US, I knew that the work context would be critical to consider.

Work can have an important impact on older adults' everyday ICT use. Workplaces provide not only access, but also formal training and informal mentoring, resources, and support. Recognizing the importance of work in influencing use, as well as the lack of literature addressing working older adults, I realized that I needed to select a cohort with large numbers of older adults still working.

Based on these factors, I decided to choose my primary participant sample from the Lucky Few birth cohort/generation (born 1929–mid-1946) (Ortman, Velkoff, & Hogan, 2014).[4] The Lucky Few generation was the youngest generation currently completely in the older adult population (age 65 or older) and, therefore, the most likely to still have participants working. In order to control more strictly for historical exposure to technology, I firmly limited the birth years of primary participants from 1936 to mid-1946, eliminating those born 1929–1935 from participating as primary participants.

Having determined who would comprise the primary participant sample, I needed to develop a sampling frame that would allow me to compare cases of older adult ICT use.

Creating a Sampling Frame

Sampling in case studies is quite different from sampling in many other types of research. In case studies, participants (be they entire organizations or single individuals) are selected on theoretical reasoning and are not intended to form a representative sample. Instead, the selection of cases seeks to understand how these cases may differ based on theoretical differences (Yin, 2009).

There are many potential variables (suspected theoretical differences) I could have designed my case sampling frame to incorporate. Notable studies have examined how ICT use by older adults is impacted by living situation (community versus institutionally dwelling, rural versus urban) (Depatie & Bigbee, 2015;

---

[4]The end of the Lucky Few generation is considered blurry, due to controversy as to when the "Baby Boom" that created the Boomer generation began. The US Census Bureau indicates that the Boomer generation begins in July 1946 (Ortman et al., 2014), so participants who were born in the first-half of 1946 were considered members of the Lucky Few generation for this study. The ends of generations are often blurry and individuals at the end/ beginning of two generations may identify with either generation. Therefore, participants throughout the study were asked, "What generation do you identify with?" Some participants stated that they identified with the "Silent Generation" (another term for the Lucky Few), some identified as being born in the same generation as famous members of the Lucky Few generation, and some participants identified as being "born in the generation before Boomers" or "not a Boomer."

Parviainen & Pirhonen, 2017; Saunders, 2004), income (Bergström, 2017), socioeconomic class (Ihm & Hsieh, 2015; Iyer & Eastman, 2006; Parviainen & Pirhonen, 2017), education (González-Oñate et al., 2015; Vroman, Arthanat, & Lysack, 2015), experience with ICTs (Jacobson et al., 2017; Lee & Coughlin, 2015; Rosenthal, 2008), disability (Opalinski, 2001), and gender (Helsper, 2010; Padilla-Góngora et al., 2017). (Studies on the impact of race and ethnicity on older adult ICT use are unfortunately seriously lacking (Normie, 2003).) Incorporating all these variables into my case sampling frame would have been impossible, given the demands in time and resources such an intensive case study format requires.

Work was an important context I wanted to capture, so work status became the first theoretical difference in my sampling frame. Older adults who are retired often are the ones that volunteer for research, due to their more flexible schedules (Bouma, 2001; van Bronswijk et al., 2002; van Bronswijk et al., 2009). This has often prevented us from understanding the experiences of working older adults. In order to ensure I had working older adults in my study, I would have to seek them out.

Gender has long been conceptualized in the literature as being an important variable impacting older adult ICT use (Helsper, 2010), so gender was selected as a second theoretical difference for my sampling frame. Gerontechnology research has found that women tend to have greater difficulties in learning to use computers in the same settings as men (Hill et al., 2008; Ng, 2008; Shoemaker, 2003) and report lower usage rates (Helsper, 2010; Kim et al., 2017). Research in the domestication sphere (i.e., age diverse) has found that men and women often use technologies for drastically different reasons: men tend to use them for escape and to avoid socialization, women use these ICTs for interaction, socialization, and engagement with others (Lie, 1996; Livingstone, 1994), and women prefer personal ICTs over ones they view as impersonal (Singh, 2001). Having established gender as an important case sampling criterion, I realized from my readings on the Lucky Few birth cohort, that while many women from this cohort worked or were still working, there were also women who had chosen to stay at home after their children were born and remained housewives (Carlson, 2008).[5] I expanded my conceptualization of work to include those that had stayed at home.

I created a sampling frame (Table 4) which sought to recruit men and women who were still working part-time, still working full time, retired, and women

[5]Many middle-class women of the Lucky Few generation worked prior to having children or being married but then chose to stay home due to societal expectations of women in this birth cohort (Carlson, 2008). Women's participation in the workplace has historically always been lower than men. Currently, 25.2 % of women aged 65–69 years and 13.8% of women aged 70–74 years have paid positions of employment. This is compared to 33.5% of men aged 65–69 and 20.8% of men aged 70–74 (US Bureau of Labor Statistics, 2018).

Table 4. Case Sampling Frame with Cases Completed.

Work Status		Gender		Totals
		Male	**Female**	
Working	Full time	Boris	Alice	4
		Harry	Belinda	
	Part-time	Fred	Jackie	3
		George		
Retired		Cleveland	Gwen	5
		Dan	June	
		Jack	Margaret	
			Nancy	
			Natalie	
Stayed at home		XX	Mary	2
			Mindy Jean	
Totals		7	10	17

who had stayed at home for the majority of their lives.[6] I defined work as "work for pay" and set full-time workforce participation to be more than 30 hours a week of paid work, and part-time participation as less than 30 hours of paid work per week. Retired older adults were those that no longer worked for pay, but once had. I sought to recruit at least one participant who met each combination of case sampling criteria. Such a framework was intended to allow for literal replications (Yin, 2009) (comparisons across cases I thought would be similar due to similar work status or gender) as well as theoretical replications (Yin, 2009) (comparisons between cases where I expected contrasting results because of different work statuses or between men and women).[7]

This sampling frame represented some issues that needed to be navigated during the study. Older adult's "self-descriptions" often did not match my own definitions when it came to work. Boris, who was self-employed in construction, tended to work about 50–60 hours a week during the summer, but often did not work during the two coldest winter months. Jackie was currently working part-time at multiple jobs while searching for additional employment. While Boris described himself as "semi-retired" (as in his youth he had often worked

[6]I remained open to recruiting men who had stayed at home for a significant period of time, having raised children, managing the household, or due to disability. However, I was unable to locate men born 1936–1946 that met these criteria.
[7]Due to the addition of two cases late in data collection to increase racial diversity, retired women outnumber retired men.

more than 80 hours a week in the summer), Jackie spoke about "working full-time" (despite the fact that her current work hours fell under 20 per week). Therefore, it became critical to discuss with the older adult during the recruitment phase as to what they meant by their self-description, asking them to provide the specific number of hours and months they worked.

I realized that many of the commonly used recruitment strategies when studying older adults (Birkland & Kaarst-Brown, 2010) would not suffice to fill this sampling frame. I could not go to a retirement community or a managed care setting and find older adults actively working full time. The strategy I eventually developed, based on snowball sampling (Goodman, 1961), sought to overcome the challenges of recruiting community-dwelling older adults.

Recruitment of Older Adults: Primary Participants

In the Gerontechnology literature, there has been a focus on recruiting older adult participants primarily from managed care residential settings or retirement communities (Birkland & Kaarst-Brown, 2010). A large population of older adults are available at these settings, and often residents are eager to be involved in projects with outside individuals for a change of pace from their typical days. While such samples are convenient for researchers, they are problematic, as only a small portion of older adults live in such settings.[8] These communities and institutions often restrict or, conversely, offer ICTs that the general older adult population may or may not have access to. For instance, Nancy's assisted living center prohibited residents from using social networking sites by blocking them, but also provided a gaming console.

Those in such captive settings, such as retirement homes and graduated care, are easy for us as researchers to access, and by that measure, we often select from them liberally. There are some very concerning ethical issues in basing much of our gerontechnological research on captive populations (Birkland & Kaarst-Brown, 2010), who may feel unable to refuse participation (Reich, 1978).

In order to overcome these sampling issues, I decided to use a referent snowballing (Goodman, 1961) recruitment method, often advised to recruit older adults over "cold calling" potential participants (Johnson & Finn, 2017). In my home institution, I approached faculty, staff, and students (as well as my neighbors) asking if they knew an older adult who met the case selection criteria and would be willing to participate. Many times, if they referred me to individuals who themselves did not meet the selection criteria, these individuals had a contact who did.

---

[8]While certain characteristics, such as decreased cognitive and physical function and advancing age are predictors of living in a residential care setting (Luppa et al., 2010), not all older adults live in such settings and those that do often do so only for a short time period (Kelly et al., 2010). Retirement communities often exclude lower-income seniors (Salkin, 2009).

This referent-based recruitment method proved useful in attracting participants who were able to meet the demanding nature of the study. With approximately 6–10 hours of the primary participant's time spent in interviews (with an open-ended 2- to 3.5-hour segment of time needing to be scheduled for each interview), additional time spent in observations in the participant's own home, and being asked to interview two to three potential friends, family members, or coworkers, this was a demanding study.[9] Compared to our stereotypes about older adults having large amounts of free time, many are busy with their leisure, community, family and (in many cases) their work lives (Bouma, 2001; Choi, Burr, Mutchler, & Caro, 2007; Eggebeen & Hogan, 1990; van Bronswijk et al., 2002).

Referents, therefore, served an important role in recruitment. Participants were not approached by a stranger, but rather a friend, family member, neighbor, or professional contact. Referents were able to describe not only the study but answer any questions the potential participant had about it at the outset, as well as vouch for me personally. Almost any person would be hesitant to allow a stranger into their home to view their technology for their own safety (such a research study would appear to be an almost perfect cover to allow criminals into your home). Additionally, referents were able to reassure potential participants that I was indeed interested in both use and non-use and that I was not trying to evangelize technology use or judge non-users.[10]

Such a referent recruitment method proved extremely successful, meeting the case sampling framework. Participants were well-aware of the steep requirements of the study from both the referent and myself, and this resulted in every older adult who enrolled in the study completing all three interviews, for a total of 17 older adult primary participants (cases). (Secondary participants were not able to be recruited for all cases, as some older adults had very small social networks.) Only one person who was referred to me (who meet the selection criteria) declined to participate.

Near the middle of data collection, I realized that while I was likely to meet my sampling frame for the study, and had achieved an educationally, work status, and income diverse sample; racial diversity was lacking. My sample identified almost entirely as white, with one participant who identified as Hispanic/White. I had serious concerns about the overwhelming whiteness represented in many gerontechnological studies (Normie, 2003) and how this further silenced older adults of color. I wanted to include voices from this often ignored a segment of the older adult population.

---

[9]Primary participants received a US$20 gift card (of their choice) at the start of each of their interviews as an incentive.

[10]Such referents also provided a check on the participant's trustworthiness and helped to provide for my own safety. I met my participants in their own homes, often venturing into unfamiliar neighborhoods and rural areas, and would be gone for an undetermined amount of time, with my friends, family, and colleagues unaware of my exact location to protect the identity of my participants.

Older individuals of color have extremely valid reasons to not participate in research. Research's abusive and racist history is not just "historical": many older adults remember these abuses when they happened (or were revealed) – in their own lifetimes. The university I was working from at the time also had a contentious relationship with the local communities of people of color. Located near one of the most impoverished sections of a racially segregated city, the university had a multi-decade history of starting community projects, promising long-term investment, obtaining research results, and quickly abandoning these projects with promises unfulfilled. As a white researcher from this institution, I would understandably be mistrusted from the start.

I decided that I needed a strategy to personally build trust with the older adults in the nearby communities. I reached out to six community organizations in my city, asking if I could join their community meetings for a period of time in order to eventually recruit participants. One community organization allowed me to do so, and I attended a series of meetings to become familiar with attendees and to build greater trust.[11] After several attendances, I was approached by two African American women. I ended up recruiting both women (Gwen and June), which greatly added to the diversity of voices.

Originally, I had determined that I would not include participants from residential care settings in my study. My original reasoning for not recruiting directly from these settings was to avoid the unnecessary burden that had been placed on these individuals by their over-selection. I also wanted to remove possible ethical concerns around consent. One of my contacts referred me to her mother, Nancy, who lived in an assisted living facility. Upon considering her inclusion, I realized that for Nancy, her home *was* the assisted living center she lived in. For many older adults, their home is their assisted living center, nursing home, or retirement community. I was not selecting her *because* she lived in assisted living, but because she *met the criteria* of the study.

As I considered the issue even further, I realized that not allowing Nancy to participate in the study simply because she had made her home in an assisted living center would be unfair to her. If my goal was to hear from the older adults themselves – to have them voice their joys, struggles, and concerns over ICT use – then I had no reason to exclude an older adult's voice because of where they had made their home. The next day I received a call from Nancy herself, asking to be included in the study and sharing a bit of her story. While I had a few participants who had contacted me (as opposed to me contacting them as a referral), she was the only participant who contacted me to argue their case to

[11]During these meetings I took part in the meeting content (only if appropriate) but I always made it clear that I was a researcher who was seeking participants. The content of these meetings was outside of the scope of my research. I was careful to not record any information about the many participants of the meeting. Upon recruiting the two individuals I recruited from this setting, I stopped attending to prevent myself from contaminating my data collection on these individuals.

be included. Quite frankly, I admired her tenacity and knew I could learn something from her case.

The data that I collected from Nancy were extremely rich and added deeply to my understanding of the ICT User Typology. As a Socializer, Nancy found herself often prevented from using the ICTs she wanted to by disability or policy (see Chapter 4). Nancy's case illustrated physical and policy concerns that I would not have captured without her inclusion.

While the primary participants for the study were selected based on the sampling criteria, secondary participants were identified and recruited based on the recommendation of the primary participant for each case.

Secondary Participants Recruitment and Sampling

For secondary participant recruitment, I asked the older adult at the center of each case (the primary participant) to identify individuals they used ICTs with (be it the phone, cell phone, computer, television, etc.). I asked them to identify one person who was of the same birth cohort as themselves (the Lucky Few generation) and one to two people who were either older (WWII/ Good Warrior Generation) or younger (Boomer, Generation X, or Millennial).[12] These relationships could be of any type: romantic, family, friendship, coworker, etc. Therefore, of the two to three people I asked them to identify, one had to be of the same birth cohort, and two were from a different birth cohort.[13]

In some cases, I ended up recruiting Lucky Few secondary participants to become primary participants in my study. This yielded very rich data, as I could understand from multiple perspectives how these individuals used ICTs in their relationships with others. This type of recruitment happened specifically for Jackie (who was recruited from Natalie's network) and Fred (who was recruited from Alice's network).

Each case involved several forms of data collection, including interviews, observations, and reviews of documentation that older adults used in their support of ICTs.

The Interviews: Collecting the Data

Data were collected over a two-year time period. Several cases were being collected at any one time; each case's data collection phase lasted approximately two months. Each case included three semi-structured interviews with the

---

[12]Due to the stipulations put on my study by my institutions' Internal Review Board (IRB), all the secondary participants had to be age 18 or older (adults). This excluded Generation Z from secondary participants, as this generation was just entering adulthood.

[13]Secondary participants received a US$10 gift card of their choice. This was typically sent through the mail, as the vast majority of these interviews were conducted on the phone.

primary participant, one semi-structured interview with each of the two to three secondary participants (if possible), observation of the primary participant's home (and workplace, if applicable) and the display of ICTs within it, and review of the documents that the older adult identified as important in their use of ICTs (if relevant).

Semi-structured interview guides were created for each primary participant interview and the secondary participant interview. The first interview with the primary participant focused on determining the ICTs used by the older adult, ICTs they had abandoned, and ICTs they wished to try. During this first interview, a set of color-coded notecards was created for each participant. Each notecard listed a single ICT: green notecards recorded ICTs the older adult was currently using, pink notecards indicated the ICTs the older adult no longer used (rejected), and yellow notecards indicated the ICTs that the older adult wished to try.

The second interview focused on understanding, through a notecard sort, how the ICTs recorded on the notecards in the first interview were used in different life contexts by the older adult (work, leisure, community/volunteering, and family life). As we addressed each of these contexts, the older adult was asked to sort out the ICTs they used in that context and speak about how they used them (and for what purposes). (New cards/ ICTs could be added.) I took pictures of the cards for each context to facilitate data analysis. This process of card sorting was repeated for the yellow (want to try or are interested in trying) and pink cards (abandoned or rejected ICTs). I encouraged the primary participant to tell me stories about the ICTs they used, rejected, or wanted to try. Once we had finished examining a life context, the notecards were returned to a pile, shuffled, and we addressed the remaining life contexts in turn with their own notecard sort.

Throughout the first two interviews, I sought to gather names of potential secondary participants. Between the second and third interview with the primary participant, I interviewed any secondary participants that had been identified that wished to participate. Secondary participants were able to be recruited for 13 of the 17 cases. The secondary participant was asked to describe their relationship with the older adult, how they had met, how often and how they communicated, with a focus on how ICTs were used in their relationship with the older adult. They were also asked to characterize their own ICT use. These interviews were often half an hour to an hour and a half in length, most often over the phone.

The third and final interview with the primary participant had two major purposes: the first was to gather further information on the relationship between the secondary and primary participants from the older adult's perspective. The second was to explore the primary participant's display of ICTs in their home and workspaces (if applicable).[14] This final interview also served as a "wrap up" and disengagement of the case. Participants were asked to share any final stories.

---

[14]If participants agreed, images of the arrangement of their ICT displays in their homes and workspaces were also taken during this visit. Any identifying information from these images was immediately removed and only anonymized images were saved. (Older adults were never included in these images.)

I also shared what I had learned from their case, but also what I was finding overall through my field work. These discussions served as a check on my findings and analysis. Often, it was during the final interview that many participants (particularly Guardians and Traditionalists) would share the most sensitive stories of ICT use and non-use. Margaret, for instance, during our third interview together shared her story of how the introduction of the computer and television to her workplace correlated with the reduction of her job responsibilities.

Interviews were transcribed immediately after an interview was conducted to allow analysis before the next interview (if appropriate). Data collection continued until the sampling frame had been filled (with the addition of two cases, June and Gwen, to increase racial diversity). At that point, data saturation had been reached, as the final cases were yielding no new nuances to the user types I had developed earlier in my data analysis.

Data Analysis

Ongoing data analysis took several forms which were often overlapping and intersecting. Analysis occurred at three points in each case: during the interviews themselves (dialogic analysis), after each interview (memoing, within case analysis, emergent themes and meanings analysis), and at the conclusion of each of the cases (between and across case analysis, emergent themes and meanings analysis).

Dialogic analysis is a reiterative process of examining a segment of text or interview, relaying that meaning to participants, and then revising that meaning until the researcher reaches shared understanding (Denzin, 2001). During the interviews, I employed active listening, through which I sought to understand the meanings presented by the primary and secondary participants.

Between interviews I utilized bracketing. Bracketing is a method of data analysis frequently used in phenomenology, which allows a researcher to separate a text from a context (Denzin, 2001). I began by gathering the stories and biographies presented in a case and bracketed elements and meanings about ICT use and non-use. I took the bracketed elements, listed them in the order in which they occurred, and then detailed how these elements were interrelated. Finally, I re-contextualized the phenomenon of technology use by incorporating the participant's biography, their history, and social environments to create themes. I then compared the themes of these stories to other stories that were similar or contrasted within the individual case I was examining. Finally, I examined these themes between the collected cases.

For these comparisons, I used many case analysis strategies outlined by Yin (2009), including explanation building (creating an explanation from a single case and comparing this explanation to others), pattern matching (creating and exploring rival explanations for similar outcomes within a case), and cross-case analysis (comparing cases to understand different outcomes). In the end, explanation building proved one of the most powerful case analysis methodologies I employed, particularly when paired with memoing (Bentz & Shapiro, 1998).

Every day during the study I reflectively memoed about the cases I was collecting (Bentz & Shapiro, 1998), creating case descriptions and listing case similarities and differences in ICT use and meanings. As I analyzed the cases, I came to understand that it was the meanings these individuals were applying to ICTs that explained the differences between cases. While coding was critical in determining the specifics to each user type and understanding their nuances, it was through memo writing about the meanings that I first came to realize that I might be seeing distinct categories or user types.

My study began with three cases: Natalie, Margaret, and Jackie, all individuals I would later determine to be Guardians. During my early memoing and bracketing process I believed that I had found a gender effect, rather than the user types I would go on to discover. When I added Alice's (an Enthusiast) case to my study, I found that her meanings and beliefs about ICTs were much different than the other women I had previously interviewed. As I was memoing in my car following an interview, exploring possible differences between Alice, Jackie, Natalie, and Margaret, I had a breakthrough: it was not gender that was making a difference between these cases, but meaning. Jackie, Natalie, and Margaret were not all similar because they were women, but rather similar because they held the same set of meanings: viewing technology as a potentially negative influence. Alice's meanings were not a reflection of a gendered interpretation of ICTs but represented a set of meanings shared by Enthusiasts.

Through memoing, I was able to build explanations of my data, developing the ICT User Typology, and then using further cases to first expand and then final cases to test the typology. While the Enthusiast and Guardian types emerged early in the study, the Practicalist and Traditionalist types emerged toward the quarter mark of data collection. Although I had interviewed a Socializer by the middle of data collection, it took more substantial memoing to separate the Socializer type from the Enthusiast type. Two thirds of the way through data collection I had discovered all five types and spent the rest of data collection determining if the following cases added any additional types to my original five (they did not), added nuances to my cases (after 13 cases the following four did not), and if they confirmed my findings (they did). I memoed each day during the almost two years of data collection, often referring to my transcripts, field notes, and analysis.

The power of the interactionist case method was not only in building the ICT User Typology, but also in its ability to recruit participants and develop rapport, which were critical to hearing the often deeply personal stories I heard from participants.

Power of the Interpretive Interactionist Case Methodology

Using the referent snowballing method helped recruit potential participants who otherwise might not have participated in a technology study due to the fear that their technology use/non-use would be negatively judged. The interpretive interactionist case methodology, because of its dialogic method which prioritizes conversation and meaning, led to the development of deep rapport with participants (Denzin, 2001). As a result of this rapport, participants shared many stories that I might not have heard otherwise, helping to develop a rich participative theory.

Recruiting Community-dwelling Working Older Adults

Recruiting community-dwelling older adults has sometimes been a challenge in the gerontechnological literature, due to older adults' busy lives (Bouma, 2001). In particular, it has often been difficult to recruit older adults who are not participants of community center programs or programs designed specifically for seniors (Birkland & Kaarst-Brown, 2010). Many older adults do not participate in such programs, and attendance of such programs may be difficult for working older adults due to scheduling.[15]

An extremely diverse sample of participants was recruited through the snowball referent methodology (Goodman, 1961). The sample included college professors, nurses, administrative assistants, electricians, contractors, retail workers, and retirees; those with extremely low incomes that depended on government assistance and those with high incomes who had retired early; those who worked as directors and VPs and those who had been front-line retail employees and administrative assistants; those that did not finish high school and even two participants with doctorates (one a PhD and one a professional doctorate). Participants lived in the inner city, urban areas, a host of suburban areas, rural towns, and in extremely rural areas. (The only aspect of diversity which fell short was racial, addressed previously in this chapter.) I cannot envision any other methodology allowing recruitment of such a diverse sample, given the unequal spread of potential recruitment sites across the geographical area.

With the exception of two of the participants whom I had met before, all the other 15 participants were total strangers to me at the outset of the study. Despite our unfamiliarity, these participants were willing to share some of their most sensitive stories with me, in part, because I had been recommended to them by someone they trusted (a friend, family member, or community contact). The referral system worked not only to recruit individuals that would be otherwise inaccessible, but it also worked in developing trust. For many participants, they regarded me as trustworthy because someone they already trusted in their lives had referred me to them. Margaret shared that she only participated in the study because her neighbor (who had been her referral contact) had reassured her I did not have a hidden "pro-technology agenda" (Margaret).

Since participants were recruited into the study by someone they trusted, they tended to trust me, as a researcher, from the start, which made them feel more

---

[15]Only two participants had participated in community center and/or senior focused programs, attesting to the fact that when we sample from such programs we are often missing many older adult experiences.

at ease. The use of the dialogic method (Denzin, 2001) also helped participants feel comfortable and helped to develop our relationship.

Developing Rapport

Mary commented at the end of our sessions together, "Oh, this is just like therapy! You are so soothing to talk to and you really listen." While active listening sought to reinterpret the meaning back to the participant (Denzin, 2001), it also resulted in participants feeling heard and we developed deep trust. Many participants shared that they had not encountered people who were interested in or open-minded when it came to their technological stories.

Older adults face many stereotypes (Glover & Branine, 1997; Longino, 2005), not only about how they think (Binstock, 2005; Cutler, 2005), but also about technology use in our societies (Cutler, 2005), including that they are slow and unwilling learners or have extremely high levels of computer anxiety (Dyck et al., 1998; Mitzner et al., 2010). Technology is stereotyped as for the young (Larsen, 1993), the young are believed to be "digital natives" (Helsper & Enyon, 2010): the valid and appropriate users of innovative new ICTs (Larsen, 1993); the old are not (Rama, De Ridder, & Bouma, 2001). Older adults often internalize these stereotypes, regardless of their skill level (Birkland, 2016).

By the end of the third interview for some participants, I was hearing painful stories which they told me that they had rarely or sometimes never shared. Margaret shared with me in our last interview that she came to see technology as slowly eroding her job responsibility. She was embarrassed to admit this, as it seemed to suggest she was overly fearful of technology in a society that worshipped it. Mindy Jean shared with me that she had little interest in using the computer and did so only to please her family members, something she would not admit to her children or husband. Natalie shared with me very personal stories about the impact of television and video games on her divorce.

Many of the participants commented that they found the study to be "fun" and "interesting" and were very keen on hearing how their case was shaping my findings. They were all eager to assist in building my theory of ICT use.

Generating a Rich Participative Theory

The notecard sort, home observations, and language checks were extremely helpful in building the ICT User Typology. The notecards, on which I had recorded the ICTs older adults were using, had abandoned and wanted to try in the future, were a powerful method of visualization. It was easy to determine an Enthusiast from their notecard sort – nearly every notecard was on display for every life context. For Practicalists, ICTs tended to only appear in one or two piles (appearing only in work or appearing only in family and community piles, for instance). Participants found using these notecards fun, with some commenting it was like a game, but it was serious enough that all participated.

By the end of the study, it was easy to distinguish the various types upon entering a person's home. Enthusiasts' would have their main living space enveloped in technologies; Traditionalists' would have their television be prominently displayed; and Socializers' would meet me at the door, cell phone in hand. Being able to observe the environment that the older adult was living in was crucial to understanding how ICTs were being used in these everyday environments.

When I showcased my ongoing results with primary participants, many had comments and suggestions, and sometimes this resulted in the generation of new stories. In developing the user type labels, I looked toward the participants' own words, but also checked my labels with my participants themselves. I wanted to avoid the often negative, derogatory, and stereotypical labels we apply toward older adult's ICT use. I do not believe that Traditionalists or Guardians should be portrayed negatively; both represent a legitimate approach toward technology which is just as valid as Practicalists, Socializers, or Enthusiasts.[16]

Despite the power of the interpretative interactionist framework, it also had several pitfalls, including difficulty with recruiting secondary participants and challenging ethical dilemmas.

Challenges with the Interpretive Interactionist Case Methodology

The delineation of the challenges I encountered in this research is a somewhat a false separation as many of these issues (recruitment, sampling, and ethical challenges) are related to one another.

Recruitment Challenges: Secondary Participants

For some of the older adult primary participants, their networks were extremely small, making identifying and recruiting secondary participants difficult.[17] Retired older adults tended to have smaller social networks than those still working. Even if network participants could be identified, it was often difficult to track them down or arrange an interview time. June identified a list of participants I could contact, however; I was unable to get in touch with most of them and those I did contact declined to participate. Jackie did not want me to contact any of her friends in her small network: she was relying on both of them as places to park her trailer when she moved out of her apartment. (Given her precarious financial situation, she did not want to endanger her relationship with either person.)

Unfortunately, there is not much that can be done to encourage greater participation of secondary participants. While a small incentive was provided for

[16]After all, many of us, who do not identify as Guardians or Traditionalists, have expressed concerns over the societal influence of technology or nostalgia for our youth.

[17]Secondary participants could only be recruited for 13 out of the 17 cases, leaving four cases with only primary participants.

the interview, this proved not to be effective.[18] Even when a primary participant, such as June, encouraged her contacts to return my calls, they often did not. While I had cultivated relationships with my primary participants by meeting them face to face, using the active listening techniques of the dialogic method, and meeting them several times over the course of the study, I was unable to cultivate the same types of relationships with secondary participants. These individuals were often not in the same area as myself (living at a distance) and to them, I was a voice on the phone, not a person who was meeting with them within their home. They were often less interested in the topic compared to primary participants as well.

Secondary participants were also likely concerned about contaminating their relationship with the primary participant. Some secondary participants were hesitant to speak about their relationship other than the basic "facts" of when and how they met, and how often they communicated. When I asked questions such as, "how would you describe your relationship with [the primary participant]?" some secondary participants were uncomfortable enough with the question to answer only in simple sentences and resisted further probing. Other secondary participants wanted to know exactly what I would share from the interview with the primary participant.

Given the amount of time I invested in contacting and arranging secondary interviews (which were often rescheduled at the last minute) compared to the ease of arranging primary participant interviews (with only two interviews rescheduled out of 51 total primary participant interviews), I would likely recommend not including secondary participants in such a case study, unless the focus of the study is relationships, or the potential data is predicted to be valuable.

Although secondary participant interviews did not provide data which contrasted or contradicted older adults' own stories of their ICT use, they did yield valuable data that served to inform the ICT User Typology. When I began to develop the typology, I realized that this was not an age-specific theory, but rather that these types were reflected in the age-diverse secondary participants as well. (These data are explored in depth in Chapter 8.) While these interviews were the most difficult to collect, and these participants the most difficult to recruit, they did add value to my study. Therefore, it is important for researchers to weigh the power of the results against the hurdles in secondary participant data collection.

Sampling Diversity Challenges

As our societies not only gray, but also become more racially and ethnically diverse (Hayes-Bautista, Hsu, Perez, & Gamboa, 2002), we should not collect all white samples (Normie, 2003). Older people of color may be more sensitive to participating in research studies because of their more intimate familiarity with

[18] A US$10 gift card of their choice.

the research abuses their population has experienced. For older adults, racist research such as the Tuskegee Experiments (active from 1932 to 1972) (Freimuth et al., 2001) is not distant history, but rather an event that happened in their lifetimes. One cannot expect a community that experienced such horrors to be receptive to researchers, no matter the study topic. Older adults of color also tend to have fewer economic resources at their disposal compared to whites (O'Brien, Wu, & Baer, 2010), impacting their ability to participate when studies require time resources. However, from an ethical and moral standpoint, it makes the voices of these older adults incredibly important to represent.

Issues of mistrust may be lessened if the researcher is a person of color and/or already a member of that community (Freimuth et al., 2001). Gwen shared that older adults in her African American community understood that their knowledge was powerful, and many felt that sharing knowledge outside the community was relinquishing power. She suggested that in future studies having an African American individual collect such information would be helpful, as a person from within the community would face less resistance not only in entering the community, but that people would be more forthcoming and willing to share their thoughts and feelings. For white researchers, it is important to establish connections with fellow researchers who identify as people of color, who can provide vital insight. However, being or involving researchers of color is simply not enough; we must ensure our studies are ethically and professionally sound (Freimuth et al., 2001). If researchers are to have any hope of building ties with these communities, we must be diligent to ensure we treat participants fairly.

Ethical Challenges

One of the greatest challenges I encountered was that the interactionist case design often led to a blurring of the relationship between researcher and participant. Over the course of the study, the high number of contact hours and my active listening style led to several participants seeing me less as a researcher and more as a friend. This was problematic and ethically concerning: I was a researcher who was going to write about their experiences, and such writing might not always involve the flattery or omission of negative details that a friendship relationship might provide. One of the most difficult cases was Natalie.

Natalie's situation was unique in the study: she was extremely socially isolated. The loss of her family in mid-life had been a devastating blow to her self-confidence. As a result, Natalie's loneliness often led her to make overtures toward friendship. She would invite me to lunch and suggest we go shopping after our interview appointments. I always politely declined and gently reminded her that I was there as a researcher.

A compromise I struck with her is that I would stay an extra half an hour to speak with her at the end of each interview, but I declined going elsewhere or ordering food. She could become quite critical when I refused further socialization, which was often followed by a cycle of her apologizing for her behavior. Throughout the study, she also often contacted me with invitations to social events. I instead researched and then suggested a list of social groups she might be interested in joining focused around her political interests and hobbies. After several months, the calls dropped off.

I also observed less than ideal living situations for several of my participants. I had met Natalie at her home during the third and final interview. She had commented in previous interviews, "You can come to my house, but I have a lot of stuff." Upon entering her home, I saw paths existed to all three downstairs doors, but otherwise, Natalie had placed items, often stacked to the ceiling, in every room of her home. I was immediately concerned for Natalie's safety and well-being and wondered if I should intervene. As my observation continued, I examined her house for several criteria: I found that she had relatively wide paths still available in her home to three exits (and many windows were also clear and could serve as an exit if necessary) and that there was no visible garbage. The home did not smell, and even though her counters were piled with items in her kitchen and bathroom, they (as well as the sinks and toilet) were nearly spotless. The floors were also freshly cleaned, and she was currently cooking using both her stove top and oven. She had running water and electricity.

After discussion with a colleague about Natalie's living situation, I decided that I would take no further action. While her situation was not ideal, Natalie did not express being unhappy with the way her home functioned or was organized, or even over the existence of so many objects in her living space. She commented that she, "had what she needed."

Jackie, another participant, had fallen on difficult times after the death of her partner. The two never married because they viewed a marriage license as a "slip of paper." Since Jackie had lived outside of the US, she did not qualify for Social Security, falling under the number of working years required to collect benefits.[19] Jackie had worked mainly hourly jobs as an administrative assistant or in retail, typically as a clerk. While her partner had left her his estate, she found she had little money left over after clearing their debts. She pieced together part-time jobs and picked up seasonal work but found it increasingly difficult to find employment. Interviewers would often comment on her age, suggesting she was unable to do the level of work required (standing, lifting, and using the checkout machines). Her work schedule was getting more difficult for her to maintain as she grew older.

Jackie had decided that she could no longer afford to maintain an apartment. She took the remainder of the estate she had inherited from her partner and bought a pull-behind popup trailer. She found a campground in the southern

---

[19]Social Security is the United States pensioner program for older adults. Typically, a person qualifies for a sum of half of their spouses (or former spouses, even in the case of death or divorce) social security benefits in lieu of collecting their own benefits. However, they must have been married for 10 years or longer, and not qualify for benefits on their own, or the amount they could collect is lower than half of their spouse's (United States Social Security Administration, 2018). While Jackie and her partner had been together for 10 years, they were not legally married.

United States that only charged US$50 a week to park a trailer, and she planned to move to the campground for nine months of the year, returning north to live on her friends' property during the summer months. She had several leads on retail jobs in the area she was moving to, convinced that US$200 a month in "rent" would be much more affordable than the US$800 she was paying for her current apartment. In the end, I realized that although I could suggest that Jackie look into social welfare programs, there was little I could do to help.

Encountering these situations in the field was challenging for me personally. What I found to be of great help was to develop a network of colleagues who were also researching older adults or other at-risk communities. Within this network, I could discuss the issues I was facing in my research, talk through my thinking, and bounce ideas and solutions off my colleagues. I was not alone, and I was not the only person experiencing ethical dilemmas and encountering challenging situations.

To any researcher working with older adults, or indeed with any at-risk population, developing such a network of colleagues is essential. In the end, unless there is a case of abuse observed, it is up to the researcher to determine their limits and when they will intervene, and how. Using such a group to talk through such issues (always keeping the identity of participants secret) can be helpful not only in determining solutions, but also in helping the researcher to deal with their own difficult emotions.

When it comes to these ethical issues highlighted when working with older adults, it is important to set boundaries, but it is also important to be flexible and understanding of your own personal limitations. Things will likely change as your study progresses, and you will likely encounter situations one would not expect. It was impossible to not feel heartbroken at Natalie's situation or to be concerned about Jackie's. While it is important to be kind to your participants, it is also important to remember to be kind to yourself.

I wish to end this book the way I started it, by acknowledging the older adult participants who made such a study possible: thank you for sharing your lives with me and for helping me to develop the ICT User Typology. The two years I spent in close contact with you will forever be a highlight of my life, professionally and personally. Thank you for your stories!

Glossary

Active Listening A form of communication where the listener reinterprets meaning back to the speaker (Denzin, 2001).

Advanced ICTs This refers to newer Information and Communication Technologies (ICTs). Advanced ICTs are understood in their current context. Eventually, advanced ICTs will be become traditional ICTs due to future technological development. Current examples of advanced ICTs include digital technologies, applications, and services.

Boomer (Generation) A generation born mid-1946–1964 (Ortman et al., 2014). The largest generation ever born in the United States (Carlson, 2009).

Birth Cohort A group of individuals who are roughly the same age and encounter historical events at roughly the same life stage (Carlson, 2009; Edmunds & Turner, 2002; Eyerman & Turner, 1998).

Bracketing A data analysis method which focuses on isolating a segment of text, understanding the meanings in that text devoid of context, and then re-contextualizing the text with appropriate contextual factors such as their background, history, and social environments (Denzin, 2001).

Case Study A methodology that allows researchers to understand rich contextual impacts on a phenomenon (Flyvberg, 2006). Case studies can use a variety of methods to obtain information about this rich context, such as interviews, observation, and focus groups (Yin, 2009).

Community Context Involvement in activities that are not purely leisure, work, or family orientated. This includes governmental/citizen activities (such as political activism) and other activities such as religious worship or belonging to a neighborhood association.

Dialogic Analysis This is a form of active listening; this interpretive interactionist methodology focuses on a reiterative process of meaning making, where the meaning shared by an individual is reinterpreted by the interviewer. This reiterative process continues until both parties reach a shared understanding (Denzin, 2001).

Direct User This is someone who themselves manipulates and uses an ICT. For example, if someone is a direct user of a computer, this means that they themselves operate the computer.

Domestication This is a theory that proposes that ICT use is complex and that contextual and social factors influence use. Domestication proposes that adoption of an ICT is a process, which involves the introduction of an ICT, how the ICT is used (including routines of use), the display of the ICT, and the meaning of the ICT to the individual and their family and friends (Silverstone, 1994, 1999, 2007; Silverstone & Haddon, 1996; Silverstone & Hirsch, 1994).

Enthusiast An individual who loves ICTs and seeks out the latest innovations. Technology is their main hobby and interest and they approach technology as a fun toy. One of the five user types in the ICT User Typology.

Family Context Individuals who are related to one another.

Gaming Refers to the act of playing games, be they board games or digital games.

Generation A birth cohort of individuals who, by nature of being born closely together, are in the same life stage when historical events occur. As a result, they develop a shared generational consciousness (Carlson, 2009; Edmunds & Turner, 2002; Eyerman & Turner, 1998). Media and technology is an important part of any generation's experience (Naab & Schwarzenegger, 2017).

Gerontechnology The study of aging and technology, or gerontology and technology. "Gerontechnological research" is empirical work that examines aging and technology use, regardless of the researcher's own identity as a Gerontechnology scholar.

Generation Z A small generation whose birth started in the year 2002. It is unclear when this generation will end.

Generation X A small generation born 1965 to 1982 (Carlson, 2009).

Good Warrior/ World War II (generation) A large generation born 1909 to 1928 (Carlson, 2009).

Guardian An individual who believes that technology use can result in negative outcomes for individuals and society. They strictly control and regulate their own ICT use. One of the five user types in the ICT User Typology.

Historical Event An occurrence that impacts individuals, be it locally, nationally, or globally. Historical events have different impacts on individuals depending upon how old they are when these events occur (Elder & Giele, 2009). Technological innovation and introduction is a historical event (Birkland & Kaarst-Brown, 2010).

Information and Communication Technologies (ICTs) Technological artifacts that are marketed to the general public that enable information sharing and/or communication between individuals and organizations. Examples include radio, television, the internet, social media, etc.

ICT User Type (or User Type) A person's main/predominant approach toward technology. This approach, or user type, impacts how they are introduced to technology, how they use it, and how they display it in their home (or work environment, if applicable).

ICT User Typology A theory which proposes that individuals, particularly older adults, can be categorized into one of five user types, each of which has a unique approach toward and view of ICTs.

ICT Form These are different types of ICT. For instance, television is a different "form" of an ICT than the telephone or radio.

ICT Version Within an ICT form, there are different versions, or updates to an ICT. For instance, LCD televisions represent a later and newer version of the television than CRT televisions.

Indirect User This is someone who does not use an ICT themselves but instructs others to use that ICT. For instance, Traditionalists tend to be indirect users of computers and the Internet as they rely on others, such as family members or friends, to complete tasks online.

Interpretive Interactionism A dialogic method that focuses on meaning making and understanding the meaning embedded in stories (Denzin, 2001).

Leisure Activities or hobbies that are done to pass time or for fun.

Life Contexts Areas of the older adult's lives. The most explored of these contexts in the gerontechnological literature has been the family, followed by leisure and work. Another important life context is community. (These life contexts are adapted from Gerontechnology (Bouma et al., 2007).)

Life Course An approach to studying individuals' lives as a series of interconnected events (Elder, 1985; Elder & Giele, 2009).

Life Event An occurrence in an individual's life which impacts their life trajectory. These can be in any area of a person's life: domestic, health, work, etc. (Elder & Giele, 2009).

Literal Replication A case study methodology in which cases are selected based upon theoretical case sampling constructs so that it is expected that the results will be similar (Yin, 2009).

Lucky Few (Generation) A small generation born 1929 to mid-1946 (Carlson, 2008). These individuals were the primary participants whose data is presented in Chapters 2 through 6.

Memoing A reflective writing that focuses on the content and experience of a research study (Bentz & Shapiro, 1998).

Millennial (Generation) A large generation born 1983 to 2001 (Carlson, 2009). These individuals have also been referred to as "digital natives," despite research demonstrating that a large diversity of skill level exists among this generation (Helsper & Enyon, 2010).

Meaning This refers to the significance, consequences, and purpose of events, experiences, and activities for individuals (Denzin, 2001).

Older Adult A person age 65 or older.

Practicalist An individual who views ICTs as tools. They are focused on the function and usability of technologies. One of the five user types in the ICT User Typology.

Primary Participant An older adult who was the center of a case. These individuals were members of the Lucky Few generation who were born 1936 to 1946.

Secondary Participant A member of the older adult's personal network, it could be a friend, coworker, family member, neighbor, etc. For each older adult primary participant, 2–3 secondary participants were interviewed, if possible.

Snowball Sampling A method of participant recruitment where future participants are recruited from existing participants' contacts (Goodman, 1961).

Socializer An individual who views ICTs as connectors. Socializers use technologies to create and maintain relationships. They often have large intergenerational families. One of the five user types in the ICT User Typology.

Technological Anxiety Fear associated with using technology (Czaja et al., 2006). Also termed technophobia. Recent research has suggested that those who have highest levels of fear associated with technology also tend to restrict and control their use (Nimrod, 2018).

Traditional ICTs This refers to older, more traditional ICTs. Traditional ICTs are understood in their current context; today's advanced ICTs will be considered traditional technologies 30 years from now due to technological development. Current examples of traditional ICTs include radio and television.

Traditionalist An individual who prefers the technology of their youth and young adulthood over later innovations. One of the five user types in the ICT User Typology.

Trajectory A series of events that a person experiences in their lives, which taken together, represent a pathway through that person's life (Elder & Giele, 2009; Fry, 2003; Giele & Elder, 1998).

References

Ang, I. (1994). Living room wars: New technologies, audience measurement, and the tactics of television consumption. In R. Silverstone & E. Hirsch (Eds.), *Technology consumption: Media and information in domestic spaces* (pp. 131–145). New York, NY: Routledge. doi:10.4324/9780203401491_chapter_8

Anger, A. (2005). e-Seniors program in Hong Kong. *Gerontechnology*, *4*(3), 176. doi:10.4017/gt.2005.04.03.010.00

Astell, A., Alm, N., Gowans, G., Ellis, M., Dye, R., & Vaughan, P. (2009). Involving older people with dementia and their careers in designing computer based support systems: Some methodological considerations. *Universal Access in the Information Society*, *8*(1), 49–58. doi:10.1007/s10209-008-0129-9

Bagnall, P., Onditi, V., Rouncefield, M., & Sommerville, I. (2006). Older people, technology, and design: A socio-technical approach. *Gerontechnology*, *5*(1), 46–50. doi:10.4017/gt.2006.05.01.005.00

Barnard, Y., Bradley, M. D., Hodgson, F., & Lloyd, A. D. (2013). Learning to use new technologies by older adults: Perceived difficulties, experimentation behaviour and usability. *Computers in Human Behavior*, *29*, 1715–1724. doi:10.1016/j.chb.2013.02.006

Barratt, J. (2007). Design for an ageing society. *Gerontechnology*, *6*(4), 188–189. doi:10.4017/gt.2007.06.04.002.00

Becker, S. A. (2004). A study of web usability for older adults seeking online health resources. *ACM Transactions on Computer-Human Interaction*, *11*(4), 387–406. doi:10.1145/1035575.1035578

Bentz, V. M., & Shapiro, J. J. (1998). *Mindful inquiry in social research*. Retrieved from https://doi.org/10.4135/9781452243412

Benz, J., Sedensky, M., Tompson, T., & Agiesta, J. (2013). *Working longer: Older Americans attitudes on work and retirement*. Retrieved from http://www.apnorc.org/PDFs/Working%20Longer/AP-NORC%20Center_Working%20Longer%20Report-FINAL.pdf

Bergström, A. (2017). Digital equality and the uptake of digital applications among seniors of different age. *Nordicom Review*, *38*(1), 79–91. doi:10.1515/nor-2017-0398

Bilefsky, D. (October 16, 2007). Top EU court backs mandatory retirement age of 65. *New York Times*. Retrieved from http://www.nytimes.com/2007/10/16/business/worldbusiness/16iht-retire.4.7913965.html

Binstock, R. H. (2005). Old-age policies, politics, and ageism. *Generations*, *29*(3), 73–78.

Birkland, J. L. H. (2016). *"It made me feel like I had Alzheimer's": Do cultural stereotypes about aging and ICTs affect older adults' ICT use identities?* Paper presented at the Tenth International Conference on Cultural Attitudes Towards Technology and Communication, London, UK.

Birkland, J. L. H., & Kaarst-Brown, M. L. (2010). *'What's so special about studying old people?': The ethical, methodological, and sampling issues surrounding the study of older adults and ICTs.* Paper presented at the Seventh international conference on Cultural Attitudes Towards Technology and Communication, Vancouver, BC, Canada. Retrieved from http://issuu.com/catac/docs/catac2010

Blit-Cohen, E., & Litwin, H. (2004). Elder participation in cyberspace: A qualitative analysis of Israeli retirees. *Journal of Aging Studies, 18*(2004), 385–398. doi:10.1016/j.jaging.2004.06.007

Blit-Cohen, E., & Litwin, H. (2005). Computer utilization in later life: Characteristics and relationship to personal well-being. *Gerontechnology, 3*(3), 138–148. doi:10.4017/gt.2005.03.03.003.00

Boechler, P. M., Foth, D., & Watchom, R. (2007). Educational technology research with older adults: Adjustments in protocol, materials, and procedures. *Educational Gerontology, 33*(3), 221–235. doi:10.1080/03601270600894089

Bouma, H. (2001). Creating adaptive technological environments. *Gerontechnology, 1*(1), 1–3. doi:10.4017/gt.2001.01.01.001.00

Bouma, H., Fozard, J. L., Bouwhuis, D. G., & Taipale, V. T. (2007). Gerontechnology in perspective. *Gerontechnology, 6*(4), 190–216. doi:10.4017/gt.2007.06.04.003.00

Bragg, J. M. (2004). Seniors will tell you what it's like to be a senior. *National Underwriter, Life, & Health, 108*(15), 19.

Breiner, J. M., Johnson, C. C., Harkness, S. S., & Koehler, C. M. (2012). What is STEM? A discussion about conceptions of STEM in education and partnerships. *School Science and Mathematics, 112(1)*, 3–11. doi:10.1111/j.1949-8594.2011.00109.x

Brophy, C., Blackler, A., & Popovic, V. (2015). *Aging and everyday technology.* Paper presented at the IASDR2015 Interplay, Brisbane, Australia. Retrieved from https://eprints.qut.edu.au/89127/1/Aging%20and%20ICT%20-%20IASDR%202015_FINAL%20.pdf

Burtless, G., & Quinn, J. F. (2001). Retirement trends and policies to encourage work among older Americans. In P. P. Budetti, R. V. Burkhauser, J. M. Gregory, & H. A. Hunt (Eds.), *Ensuring health and income security for an aging workforce* (pp. 375–415). Kalamazoo, MI: W.E. Upjohn Institute for Employment Research. doi:10.17848/9780880994668.ch18

Buse, C. E. (2009). When you retire, does everything become leisure? Information and communication technology use and the work/ leisure boundary in retirement. *New Media & Society, 11(7)*, 1143–1161. doi:10.1177/1461444809342052

Carlson, E. (2008). *The Lucky Few: Between the Greatest Generation and the Baby Boom.* New York, NY: Springer.

Carlson, E. (2009). 20th-century: US generations. *Population Reference Bureau, 64*(1), 1–18

Charness, N. (2006). Work, older workers, and technology. *Generations, 30*(2), 25–30.

Charness, N., & Boot, W. R. (2016). Technology, gaming, and social networking. *Handbook of the Psychology of Aging* (8th Ed., pp. 389–407). Retrieved from https://doi.org/10.1016/B978-0-12-411469-2.00020-0

Charness, N., Kelly, C. L., Bosman, E. A., & Melvin, M. (2001). Word-processing training and re-training: Effects of adult age, experience, and interface. *Psychology and Aging*, *16*(1), 110–127. doi:10.1037//0882-7974.16.1.110

Charness, N., Schumann, C. E., & Boritz, G. M. (1992). Training older adults in word processing: Effects of age, training technique, and computer anxiety. *International Journal of Technology and Aging*, *5*(1), 79–106.

Chatterji, S., Byles, J., Cutler, D., Seeman, T., & Verdes, E. (2015). Health, functioning, and disability in older adults-present status and future implications. *The Lancet*, *385*, 563–575. doi:10.1016/S0140-6736(14)61462-8

Chen, Y.-R. R., & Schulz, P. J. (2016). The effect of information communication technology interventions on reducing social isolation in the elderly: A systematic review. *Journal of Medical Internet Research*, *18*(1), e18. doi:10.2196/jmir.4596

Choi, N. G., Burr, J. A., Mutchler, J. E., & Caro, F. G. (2007). Formal and informal volunteer activity and spousal caregiving among older adults. *Research on Aging*, *29*(2), 99–124. doi:10.1177/0164027506296759

Clark, D. J. (2001). Older adults living through and with their computers. *Computers, Informatics, Nursing*, *2002*(May/June), 117–124. doi:10.1097/00024665-200205000-00012

Cockburn, C. (1994). The circuit of technology: Gender, identity, and power. In R. Silverstone & E. Hirsch (Eds.), *Consuming technologies: Media and information in domestic spaces* (pp. 32–47). New York, NY: Routledge. doi:10.4324/9780203401491_chapter_2

Coeckelbergh, M. (2018). Technology and the good society: A polemical essay on social ontology, political principles, and responsibility for technology. *Technology in Society*, *52*, 4–9. doi:10.1016/j.techsoc.2016.12.002

Coelho, J., & Duarte, C. (2016). A literature survey on older adults' use of social network services and social applications. *Computers in Human Behavior*, *58*, 187–205. doi:10.1016/j.chb.2015.12.053

Coleman, L. J., Hladikova, M., & Savelyeva, M. (2006). The baby boomer market. *Journal of Targeting, Measurement, and Analysis for Marketing*, *14*(3), 191–209. doi:10.1057/palgrave.jt.5740181

Cotten, S. R., Ford, G., Ford, S., & Hale, T. M. (2012). Internet use and depression among older adults. *Computers in Human Behavior*, *28(2)*, 496–499. doi:10.1016/j.chb.2011.10.021

Cutler, S. J. (2005). Ageism and technology. *Generations*, *29*(3), 67–72.

Czaja, S. J. (2016). Long-term care services and support systems for older adults: The role of technology. *American Psychologist*, *71*(4), 294–301. doi:10.1037/a0040258

Czaja, S. J., Boot, W. R., Charness, N., Rogers, W. A., & Sharit, J. (2018). Improving social support for older adults through technology: Findings from the PRISM randomized controlled trial. *Gerontologist*, *3*(8). doi:10.1093/geront/gnw249

Czaja, S. J., Charness, N., Fisk, A. D., Hertzog, C., Nair, S. N., Rogers, W. A., & Sharit, J. (2006). Factors predicting the use of technology: Findings from the Center for Research and Education on Aging and Technology Enhancement (CREATE). *Psychology and Aging*, *21*(2), 333–352. doi:10.1037/0882-7974.21.2.333

Czaja, S. J., & Lee, C. C. (2007). The impact of aging on access to technology. *Universal Access in the Information Society*, *5*(4), 341–349. doi:10.1007/s10209-006-0060-x

Czaja, S. J., & Sharit, J. (1988). Ability–performance relationships as a function of age and task experience for a data entry task. *Journal of Experimental Psychology: Applied*, *4*(4), 332–351. doi:10.1037/1076-898X.4.4.332

Czaja, S. J., & Sharit, J. (1998). Age differences in attitudes toward computers. *Journal of Gerontology: Psychological Sciences*, *53B*(5), 329–340. doi:10.1093/geronb/53b.5.p329

Czaja, S. J., Sharit, J., Charness, N., Fisk, A. D., & Rogers, W. A. (2001). The Center for Research and Eduation on Aging and Technology Enhancement (CREATE): A program to enhance technology for older adults. *Gerontechnology*, *1*(1), 50–59. doi:10.4017/gt.2001.01.01.005.00

Davis, F. D., Bagozzi, R. P., & Warshaw, P. R. (1989). User acceptance of computer technology: A comparison of two theoretical models. *Management Science*, *35 (8)*, 982–1003. doi:10.1287/mnsc.35.8.982

De Schutter, B., & Malliet, S. (2014). The older player of digital games: A classification based on perceived need satisfaction. *Communications*, *39*, 67–88. doi:10.1515/commun-2014-0005

De Shutter, B., Brown, J. A., & Abeele, V. V. (2015). The domestication of digital games in the lives of older adults. *New Media & Society*, *17*(7), 1170–1186. doi:10.1177/1461444814522945

De Shutter, B., & Malliet, S. (2014). The older player of digital games: A classification based on perceived need satisfaction. *Communications*, *39*, 67–88. doi:10.1515/commun-2014-0005

DeLong, D. W. (2004). *Lost knowledge: Confronting the threat of an aging workforce*. New York, NY: Oxford University Press.

Denzin, N. K. (2001). *Interpretive interactionism*. Retrieved from https://doi.org/10.4135/9781412984591

Depatie, A., & Bigbee, J. L. (2015). Rural older adult readiness to adopt mobile health technology: A descriptive study. *Online Journal of Rural Nursing and Health Care*, *15*(1), 150–184. doi:10.14574/ojrnhc.v15i1.346

Dickinson, A., & Dewsbury, G. (2006). Designing computer technologies with older people. *Gerontechnology*, *5*(1), 1–3. doi:10.4017/gt.2006.05.01.001.00

Dickinson, A., Goodman, J., Syme, A., Eisma, R., Tiwari, L., Mival, O., & Newell, A. (2004). Domesticating technology: In-home requirements gathering with frail older people. In C. Stephandis (Ed.), *Universal access in HCI: Inclusive design in the information society* (Vol. 4, pp. 827–831). Mahwah, NJ: Lawrence Erlbaum Associates.

Directorate-General for Internal Policies. (2015). *Encouraging STEM studies for the labour market: Labour market situation and comparison of practices targeted at young people in different member states*. (PE 542.199). Policy Department A: Economic and Scientific Policy. Retrieved from http://www.europarl.europa.eu/RegData/etudes/STUD/2015/542199/IPOL_STU(2015)542199_EN.pdf

Dunnett, C. W. (1998). Senior citizens tracking technology. *Educational Media International*, *35*(1), 9–12. doi:10.1080/0952398980350104

Dyck, J., Gee, N., & Smither, J. A. (1998). The changing construct of computer anxiety for younger and older adults. *Computers in Human Behavior*, *14*(1), 61–77. doi:10.1016/s0747-5632(97)00032-0

Dyck, J., & Smither, J. A. (1996). Older adults' acquisition of world processing: The contribution of cognitive abilities and computer anxiety. *Computers in Human Behavior*, *12*(1), 107–119. doi:10.1016/0747-5632(95)00022-4

Eastman, J. K., & Iyer, R. (2004). The elderly's uses and attitudes towards the Internet. *Journal of Consumer Marketing*, *21*(3), 208–220. doi:10.1108/07363760410534759

Eaton, J., & Salari, S. (2005). Environments for lifelong learning in senior centers. *Educational Gerontology*, *31(6)*, 461–480. doi:10.1080/03601270590928189

Edmunds, J., & Turner, B. S. (2002). *Generations, culture and society*. Philadelphia, PA: Open University Press.

Eggebeen, D. J., & Hogan, D. P. (1990). Giving between generations in American families. *Human Nature*, *1*(3), 211–232. doi:10.1007/bf02733984

Elder, G. H. Jr. (1985). *Life course dynamics: Trajectories and transitions*. Ithaca, NY: Cornell University Press.

Elder, G. H. Jr., & Giele, J. Z. (2009). Life course studies: An evolving field. In G. H. Elder, Jr. & J. Z. Giele (Eds.), *The craft of life course research*. New York, NY: Guilford Press.

Eyerman, R., & Turner, B. S. (1998). Outline of a theory of generations. *European Journal of Social Theory*, *1(1)*, 91–106. doi:10.1177/136843198001001007

Eynon, R., & Helsper, E. (2010). Adults learning online: Digital choice and/or digital exclusion? *New Media & Society*, *13*(4), 534–551. doi:10.1177/1461444810374789

Fang, M. L., Coatta, K., Badger, M., Wu, S., Easton, M., Nygard, L., ... Sixsmith, A. (2017). Informing understandings of mild cognitive impairment for older adults: Implications from a scoping review. *Journal of Applied Gerontology*, *36*(7), 808–839. doi:10.1177/0733464815589987

Federal Communications Commission. (2018). *Lifeline program for low-income consumers*. Retrieved from https://www.fcc.gov/general/lifeline-program-low-income-consumers

Fischer, S. H., David, D., Crotty, B. H., Dierks, M., & Safran, C. (2014). Acceptance and use of health information technology by community-dwelling elders. *International Journal of Medical Informatics*, *83*(9), 624–635. doi:10.1016/j.ijmedinf.2014.06.005

Flynn, B. (2003). Geography of the digital hearth. *Information, Communication, & Society*, *6*(4), 551–576. doi:10.1080/1369118032000163259

Flyvberg, B. (2006). Five misunderstandings about case-study research. *Qualitative Inquiry*, *12*(2), 219–245. doi:10.1177/1077800405284363

Fozard, J. L., Rietsema, J., Bourma, H., & Graafmans, J. A. M. (2000). Gerontechnology: Creating enabling environments for the challenges and opportunities of aging. *Educational Gerontology*, *26(4)*, 331–344. doi:10.1080/036012700407820

Freedman, V. A., Calkins, M., & Haitsma, K. (2005). An exploratory study of barriers to implementing technology in US residential long-term care settings. *Gerontechnology*, *4*(2), 86–100. doi:10.4017/gt.2005.04.02.004.00

Freeman, I. C. (2005). Advocacy in aging: Notes for the next generation. *Families in Society*, *86*(3), 419–423. doi:10.1606/1044-3894.3440

Freimuth, V. S., Quinn, S. C., Thomas, S. B., Colea, G., Zook, E., & Duncan, T. (2001). African Americans' views on research and the Tuskegee Syphilis Study. *Social Science and Medicine, 52(5)*, 797–808. doi:10.1016/S0277-9536(00)00178-7

Friemel, T. N. (2016). The digital divide has grown old: Determinants of a digital divide among seniors. *New Media & Society, 18*(2), 313–331. doi:10.1177/1461444814538648

Fry, C. L. (2003). The life course as a cultural construct. In R. A. Settersten, Jr. (Ed.), *Invitation to the life course: Toward new understanding of later life* (pp. 269–294). Amityville, NY: Baywood.

Gell, N. M., Rosenberg, D. E., Demiris, G., LaCroix, A. Z., & Patel, K. V. (2015). Patterns of technology use among older adults with and without disabilities. *Gerontologist, 55*(3), 412–421 doi:10.1093/geront/gnt166

Giele, J. Z., & Elder, G. H. Jr. (1998). Life course research: Development of a field. In J. Z. Giele & G. H. Elder, Jr. (Eds.), *Methods of life course research: Qualitative and quantitative approaches* (pp. 5–27). Thousand Oaks, CA: Sage. doi:10.4135/9781483348919.n1

Gilleard, C., & Higgs, P. (2008). Internet use and the digital divide in the English longitudinal study of ageing. *European Journal of Aging, 5*(3), 233–239. doi:10.1007/s10433-008-0083-7

Gitlin, L. (1995). Why older people accept or reject assistive technology. *Generations, 19*(1), 41–47.

Glover, I., & Branine, M. (1997). Ageism and the labour process: Towards a research agenda. *Personnel Review, 26*(4), 274–292. doi:10.1108/00483489710172079

Golant, S. M. (2017). A theoretical model to explain the smart technology adoption behaviors of elder consumers (Elderadopt). *Journal of Aging Studies, 42*, 56–73. doi:10.1016/j.jaging.2017.07.003

González-Oñate, C., Fanjul-Peyró, C., & Cabezuelo-Lorenzo, F. (2015). Use, consumption and knowledge of new technologies by elderly people in France, United Kingdom and Spain. *Comunicar, 45*(XXIII), 19–27. doi:10.3916/C45-2015-02

Gonzalez, H. B., & Kuenzi, J. J. (2012). *Science, Technology, Engineering, and Mathematics (STEM) education: A primer*. Retrieved from https://fas.org/sgp/crs/misc/R42642.pdf

Goodman, L. A. (1961). Snowball sampling. *Annals of Mathematical Statistics, 32*, 148–170. doi:10.1214/aoms/1177705148

Graafmans, J. A. M., & Brouwers, T. (1989). Gerontechnology, the modelling of normal aging. *Proceedings of the Human Factors Society 33rd Annual Meeting*, 187–190. doi:10.1177/154193128903300308

Habib, L., & Cornford, T. (2002). Computers in the home: Domestication and gender. *Information, Technology, & People, 15*(2), 159–174. doi:10.1108/09593840210430589

Haddon, L. (2000). Social exclusion and information and communication technologies. *New Media & Society, 2*(4), 387–406. doi:10.1177/1461444800002004001

Haddon, L. (2007). Roger Silverstone's legacies: Domestication. *New Media & Society, 9*(1), 25–32. doi:10.1177/1461444807075201

Hauk, N., Hüffmeier, J., & Krumma, S. (2018). Ready to be a silver surfer? A meta-analysis on the relationship between chronological age and technology acceptance. *Computers in Human Behavior*, *84*, 304–319. doi:10.1016/j.chb.2018.01.020

Hayes-Bautista, D. E., Hsu, P., Perez, A., & Gamboa, C. (2002). The 'browning' of the graying of America: Diversity in the elderly population and policy implications. *Generations*, *26*(3), 15–24.

Heart, T., & Kalderon, E. (2013). Older adults: Are they ready to adopt health-related ICT? *International Journal of Medical Informatics*, *82*(11), e209–e231. doi:10.1016/j.ijmedinf.2011.03.002

Hedge, J. W., Borman, W. C., & Lammlein, S. E. (2006). *The aging workforce: Realities, myths, and implications for organizations*. Retrieved from https://doi.org/10.1037/11325-000

Heinz, M., Martin, P., Margrett, J. A., Yearns, M., Franke, W., Yang, H., … Chang, C. K. (2013). Perceptions of technology among older adults. *Journal of Gerontological Nursing*, *39*(1), 42–51. doi:10.3928/00989134-20121204-04

Helsper, E. J. (2010). Gendered Internet use across generations and life stages. *Communication Research*, *37*(3), 352–374. doi:10.1177/0093650209356439

Helsper, E. J., & Enyon, R. (2010). Digital natives: Where is the evidence? *British Educational Research Journal*, *36*(3), 503–520. doi:10.1080/01411920902989227

Hill, R., Betts, L. R., & Gardner, S. E. (2015). Older adults' experiences and perceptions of digital technology: (Dis)empowerment, wellbeing, and inclusion. *Computers in Human Behavior*, *58*, 415–423. doi:10.1016/j.chb.2015.01.062

Hill, R., Beynon-Davies, P., & Williams, M. (2008). Older people and internet engagement: Acknowledging social moderators of internet adoption, access, and use. *Information Technology and People*, *21*(3), 244–266. doi:10.1108/09593840810896019

Hilt, M. L., & Lipschultz, J. H. (2004). Elderly Americans and the internet: E-mail, TV news, information and entertainment websites. *Educational Gerontology*, *30*(1), 57–72. doi:10.1080/03601270490249166

Hogan, M. (2005). Technophobia amongst older adults in Ireland. *Irish Journal of Management*, *27*(1), 57–77.

Hough, M., & Kobylanski, A. (2009). Increasing elder consumer interactions with information technology. *Journal of Consumer Marketing*, *26*(1), 39–48. doi:10.1108/07363760910927037

Ihm, J., & Hsieh, Y. P. (2015). The implications of information and communication technology use for the social wellbeing of older adults. *Information, Communication, and Society*, *18*(10), 1123–1138. doi:10.1080/1369118X.2015.1019912

Iyer, R., & Eastman, J. K. (2006). The elderly and their attitudes toward the internet: The impact on internet use, purchase, and comparison shopping. *Journal of Marketing Theory and Practice*, *14*(1), 57–67. doi:10.2753/mtp1069-6679140104

Jacobson, J., Lin, C. Z., & McEwen. (2017). Aging with technology: Seniors and mobile connections. *Canadian Journal of Communication*, *42(2)*, 331–357. doi:10.22230/cjc2017v42n2a3221

Jay, G. M., & Willis, S. L. (1992). Influence of direct computer experience on older aults' attitudes toward computers. *Journal of Gerontology: Psychological Sciences*, *47*(4), 250–257. doi:0.1093/geronj/47.4.p250

Johnson, J., & Finn, K. (2017). Working with older adults. *Designing User Interfaces for an Aging Population* (pp. 159–180): Burlington, MA: Morgan Kaufman. doi:10.1016/b978-0-12-804467-4.00010-4

Jung, E. H., Walden, J., Johsnon, A. C., & Sundar, S. S. (2017). Social networking in the aging context: Why older adults use or avoid Facebook. *Telematrics and Informatics, 34*(7), 1071–1080. doi:10.1016/j.tele.2017.04.015

Kaarst-Brown, M. L. (1995). *A theory of information technology cultures: Magic dragons, wizards and archetypal patterns.* Ph.D. thesis, York University, Toronto, ON, Canada.

Kaarst-Brown, M. L. (2005). Understanding an organization's view of the CIO: The role of assumptions about IT. *MIS Quarterly Executive, 4*(2), 287–301.

Kaarst-Brown, M. L., & Robey, D. (1999). More on myth, magic and metaphor: Cultural insights into the management of information technology in organizations. *Information, Technology, & People, 12*(2), 192–218. doi:10.1108/09593849910267251

Kashchuk, I., & Ivankina, L. (2015). Marketing approach to the research of older adults' well-being. *Procedia – Social and Behavioral Sciences, 214*, 911–915. doi:10.1016/j.sbspro.2015.11.752

Kelly, A., Conell-Price, J., Covinsky, K., Cenzer, I. S., Chang, A., Boscardin, W. J., & Smith, A. K. (2010). Length of stay for older adults residing in nursing homes at the end of life. *Journal of the American Geriatrics Society, 58*(9), 1701–1706. doi:10.1111/j.1532-5415.2010.03005.x

Khvorostianov, N., Elias, N., & Nimrod, G. (2011). 'Without it I am nothing': The Internet in the lives of older immigrants. *New Media & Society, 14*(4), 583–599. doi:10.1177/1461444811142159

Kim, J., Lee, H. Y., Christensen, M. C., & Merighi, J. R. (2017). Technology access and use, and their associations with social engagement among older adults: Do women and men differ? *Journals of Gerontology: Social Sciences, 72*(5), 836–845. doi:10.1093/geronb/gbw123

Kinsella, K., & Velkoff, V. (2001). *An aging world.* (series P95/01-1). Washington, DC: US Census Bureau. Retrieved from https://www.cdc.gov/mmwr/preview/mmwrhtml/mm5206a2.htm

Kuenzi, J. J. (2008). *Science, Technology, Engineering, and Mathematics (STEM) education: Background, federal policy, and legislative action.* (35) Retrieved from http://digitalcommons.unl.edu/crsdocs/35

Lagana, L. (2008). Enhancing the attitudes and self-efficacy of older adults toward computers and the internet: Results of a pilot study. *Educational Gerontology, 34*(9), 831–843. doi:10.1080/03601270802243713

Laguna, K., & Babcock, R. (1997). Computer anxiety in young and older adults: Implications for human-computer interactions in older populations. *Computers in Human Behavior, 13*(3), 317–326. doi:10.1016/s0747-5632(97)00012-5

Lam, J. C. Y., & Lee, M. K. O. (2006). Digital inclusiveness – Longitudinal study of Internet adoption by older adults. *Journal of Management Information Systems, 22*(4), 177–206. doi:10.2753/mis0742-1222220407

Larsen, R. S. (1993). *Technological generations and the spread of social definition of new technologies.* Ph.D. thesis, University of Oregon, Eugene, Oregon, USA.

Lee, C., & Coughlin, J. F. (2015). Perspactive: Older adults' adoption of technology: An integrated approach to identifying determinants and barriers. *Journal of Product Innovation Management*, *32*(5), 747–759. doi:10.1111/jpim.12176

Lenstra, N. (2017). The community-based information infrastructure of older adult digital learning. *Nordicom Review*, *38(s1)*, 65–77. doi:10.1515/nor-2017-0401

Lie, M. (1996). Gender in the image of technology. In M. Lie & K. H. Sørensen (Eds.), *Making technology our own? Domesticating technology into everyday life* (pp. 201–223). Boston, MA: Scandinavian University Press.

Lie, M., & Sørensen, K. H. (1996). Making technology our own? Domesticating technology in everyday life. In M. Lie & K. H. Sørensen (Eds.), *Making technology our own? Domesticating technology in everyday life* (pp. 1–30). Boston, MA: Scandinavian University Press.

Livingstone, S. (1994). The meaning of domestic technologies: A personal construct analysis of familial gender relations. In R. Silverstone & E. Hirsch (Eds.), *Consuming technologies: Media and information in domestic spaces.* (pp. 113–130). New York, NY: Routledge.

Longino, C. E. (2005). The future of ageism: Baby boomers at the doorstep. *Generations*, *29*(3), 79–83.

Luppa, M., Luck, T., Weyerer, S., König, H. h., Brähler, E., & Riedel-Heller, S. (2010). Prediction of institutionalization in the elderly: A systematic review. *Age and Ageing*, *39(1)*, 31–38. doi:10.1093/ageing/afp202

Majumder, S., Aghayi, E., Noferesti, M., Memarzadeh-Tehran, H., Mondal, T., Pang, Z., & Deen, M. J. (2017). Smart homes for elderly healthcare – Recent advances and research challenges. *Sensors*, *17(11)*, 1–32. doi:10.3390/s17112496

Mandel, L. (1967). The computer girls. *Cosmopolitan*. Retrieved from http://thecomputerboys.com/wp-content/uploads/2011/06/cosmopolitan-april-1967-1-large.jpg

Marrero, M. E., Gunning, A., & Germain-Williams, T. (2014). What is STEM education? *Global Education Review*, *1*(4), 1–6.

McMellon, C., & Schiffman, L. (2002). Cybersenior empowerment: How some older individuals are taking control of their lives. *Journal of Applied Gerontology*, *21*(2), 157–175. doi:10.1177/07364802021002002

Melenhorst, A., Rogers, W. A., & Bouwhuis, D. G. (2006). Older adults' motivated choice for technological innovation: Evidence for benefit-driven society. *Psychology and Aging*, *21*(1), 190–195. doi:10.1037/0882-7974.21.1.190

Microsoft. (2017). *Why don't European girls like science or technology?* Retrieved from https://news.microsoft.com/europe/features/dont-european-girls-like-science-technology/#W3R2xCWJmcHeAkSH.99

Millward, P. (2003). The "grey digital divide": Perception, exclusion and barriers of access to the internet for older people. *First Monday*, *8*(7). doi:10.5210/fm.v8i7.1066

Mitzner, T. L., Boron, J. B., Fausset, C. B., Adam, A. E., Charness, N., Czaja, S. J., … Sharit, J. (2010). Older adults talk technology: Technology usage and attitudes. *Computers in Human Behavior*, *26*(6). doi:10.1016/j.chb.2010.06.020

Morrison, R. (2015). Silver surfers search for gold: A study into the online information-seeking skills of those over fifty. *Ageing International*, *20*, 300–310. doi:10.1007/s12126-015-9224-4

Mortenson, W. B., Sixsmith, A., & Woolrych, A. (2015). The power(s) of observation: Theoretical perspectives on surveillance technologies and older people. *Ageing & Society*, *35*(3), 512–530. doi:10.1017/S0144686X13000846

Mostaghel, R. (2016). Innovation and technology for the elderly: Systematic literature review. *Journal of Business Research*, *69*(11), 4896–4900. doi:10.1016/j.jbusres.2016.04.049

Naab, T., & Schwarzenegger, C. (2017). Why ageing is more important than being old: Understanding the elderly in a mediatized world. *Nordicom Review*, *38*(1), 93–107. doi:10.1515/nor-2017-0400

Ng, C.-h. (2008). Motivation among older adults in learning computing technologies: A grounded model. *Educational Gerontology*, *34*(1), 1–14. doi:10.1080/03601270701763845

Nimrod, G. (2018). Technophobia among older Internet users. *Educational Gerontology*, *44*(2–3), 148–162. doi:10.1080/03601277.2018.1428145

Normie, L. (2003). Older people, computers, and ethnicity: An academic research backwater? *Gerontechnology*, *2*(4), 285–288. doi:10.4017/gt.2003.02.04.001.00

O'Brien, E., Wu, K. B., & Baer, D. (2010). *Older Americans in poverty: A snapshot* (2010–03). Washington, DC: AARP Public Policy Institute. Retrieved from https://assets.aarp.org/rgcenter/ppi/econ-sec/2010-03-poverty.pdf

Okonji, P., Lhussier, M., Bailey, C., & Cattan, M. (2015). Internet use: Perceptions and experiences of visually impaired older adults. *Journal of Social Inclusion*, *6*(1), 120–145.

Opalinski, L. (2001). Older adults and the digital divide: Amassing results of a web-based survey. *Journal of Technology for Human Services*, *18*(3/4), 203–221. doi:10.1300/j017v18n03_13

Ortman, J. M., Velkoff, V. A., & Hogan, H. (2014). *An aging nation: The older population in the United States*. (P25-1140). Washington, DC: United States Census Bureau. Retrieved from https://www.census.gov/library/publications/2014/demo/p25-1140.html

Padilla-Góngora, D., López-Liria, R., del Pilar Díaz-López, M., Aguilar-Parra, J. M., Estela Vargas-Muñoz, M., & Rocamora-Pérezb, P. (2017). Habits of the elderly regarding access to the new information and communication technologies. *Procedia – Social and Behavioral Sciences*, *237*, 1412–1417. doi:10.1016/j.sbspro.2017.02.206

Parviainen, J., & Pirhonen, J. (2017). Vulnerable bodies in human–robot interactions: Embodiment as ethical issue in robot care for the elderly. *Transformations*, *29*, 104–115.

Paul, G., & Stegbauer, C. (2005). Is the digital divide between young and elderly people increasing? *First Monday*, *10*(10). doi:10.5210/fm.v10i10.1286

Peral-Peral, B., Arenas-Gaitán, J., & Villarejo-Ramos, A. F. (2015). From digital divide to psycho-digital divide: Elders and online social networks. *Comunicar*, *23*(45), 57–64. doi:10.3916/C45-2015-06

Petrovčič, A., Vehovar, V., & Dolnicar, V. (2016). Landline and mobile phone communication in social companionship networks of older adults: An empirical investigation in Slovenia. *Technology in Society*, *45*, 91–102. doi:10.1016/j.techsoc.2016.02.007

Pick, J. B., Sarkar, A., & Johnson, J. (2015). United States digital divide: State level analysis of spatial clustering and multivariate determinants of ICT utilization. *Socio-Economic Planning Sciences*, *49*(C), 16–32. doi:10.1016/j.seps.2014.09.001

Plawecki, H. M., & Plawecki, L. H. (2015). The emerging Baby Boomer health care crisis. *Journal of Gerontological Nursing*, *41*(11), 3–5. doi:10.3928/00989134-20151015-22

Rama, M. D., De Ridder, H., & Bouma, H. (2001). Technology generation and age in using layered user interfaces. *Gerontechnology*, *1*(1), 25–40. doi:10.4017/gt.2001.01.01.003.00

Reich, W. T. (1978). Ethical issues related to research involving elderly subjects. *The Gerontologist*, *18*(4), 326–337. doi:10.1093/geront/18.4.326

Reisenwitz, T., & Iyer, R. (2007). A comparison of younger and older baby boomers: Investigating the viability of cohort segmentation. *Journal of Consumer Marketing*, *24*(4), 202–213. doi:10.1108/07363760710755995

Reisenwitz, T., Iyer, R., Kuhlmeier, D., & Eastman, J. K. (2007). The elderly's internet usage: An updated look. *Journal of Consumer Marketing*, *24*(7), 406–418. doi:10.1108/07363760710834825

Righi, V., Sayago, S., & Blat, J. (2017). When we talk about older people in HCI, who are we talking about? Towards a 'turn to community' in the design of technologies for a growing ageing population. *International Journal of Human-Computer Studies*, *108*, 15–31. doi:10.1016/j.ijhcs.2017.06.005

Rogers, E. M. (1962). *Diffusion of innovations*. New York, NY: The Free Press.

Rogers, E. M. (2003). *Diffusion of innovations* (5th ed.). New York, NY: Free Press.

Rogers, E. M., & Shoemaker, F. F. (1971). *Communication of innovations: A cross-cultural approach*. New York, NY: The Free Press.

Rogers, W. A., & Mitzner, T. L. (2017). Envisioning the future for older adults: Autonomy, health, well-being, and social connectedness with technology support. *Futures*, *87*, 133–139. doi:10.1016/j.futures.2016.07.002

Rosenthal, R. (2008). Older computer-literate women: Their motivations, obstacles, and paths to success. *Educational Gerontology*, *34*(7), 610–626. doi:10.1080/03601270801949427

Rosowsky, E. (2005). Ageism and professional training in aging: Who will be there to help. *Generations*, *29*(3), 55–58.

Rowe, M., & Trejos, J. (2017). The dark destiny of safety. *Politica Northwestern*, *1*, 30–33. doi:10.21985/N23S3H

Russell, C. (1998). The haves and the want-nots. *American Demographics*, *20*(4), 10–12.

Salkin, P. (2009). A quiet crisis in America: Meeting the affordable housing needs of the invisible low-income healthy seniors. *Georgetown Journal on Poverty Law Policy*, *16*(2), 285–314.

Saunders, E. J. (2004). Maximizing computer use among the elderly in rural senior centers. *Educational Gerontology*, *30*, 573–585. doi:10.1080/03601270490466967

Schulz, R., Wahl, H.-W., Matthews, J. T., Dabbs, A. D. V., Beach, S. R., & Czaja, S. J. (2015). Advancing the aging and technology agenda in Gerontology. *Gerontologist*, *55*(5), 724–734. doi:10.1093/geront/gnu071

Scott, J., & Alwin, D. (1998). Retrospective versus prospective measurement of life histories in longitudinal research. In J. Z. Giele & G. H. Elder, Jr. (Eds.),

Methods of life course research: Qualitative and quantitative approaches (pp. 98–127). Thousand Oaks, CA: Sage. doi:10.4135/9781483348919.n5

Shoemaker, S. (2003). Acquisition of computer skills by older users: A mixed methods study. *Research Strategies, 19*, 165–180. doi:10.1016/j.resstr.2005.01.003

Silverstone, R. (1994). *Television and everyday life.* New York, NY: Routledge. doi:10.4324/9780203358948

Silverstone, R. (1999). *Why study the media?* Thousand Oaks, CA: Sage. doi:10.4135/9781446219461

Silverstone, R. (2007). *Media and morality: On the rise of Mediapolis.* Malden, MA: Polity Press.

Silverstone, R., & Haddon, L. (1996). Design and the domestication of information and communciation technologies: Technical change and everyday life. In R. Mansell & R. Silverstone (Eds.), *Communication by design: The politics of information and communication technologies* (pp. 44–74). New York, NY: Oxford University Press.

Silverstone, R., & Hirsch, E. (1992). *Consuming technologies: Media and information in domestic spaces.* New York, NY: Routledge. doi:10.4324/9780203401491

Silverstone, R., & Hirsch, E. (1994). Introduction. In R. Silverstone & E. Hirsch (Eds.), *Consuming technologies: Media and information in domestic spaces* (pp. 1–11). New York, NY: Routledge.

Silverstone, R., Hirsch, E., & Morley, D. (1994). Information and communication technologies and the moral economy of the household. In R. Silverstone & E. Hirsch (Eds.), *Consuming technologies: Media and information in domestic spaces* (pp. 15–31). New York, NY: Routledge. doi:10.4324/9780203401491_chapter_1

Singh, S. (2001). Gender and the use of the internet at home. *New Media & Society, 3(4)*, 395–415. doi:10.1177/14614448010030040001

Slegers, K., van Boxtel, M. P. J., & Jolles, J. (2007). The effects of computer training and internet usage on the use of everyday technology by older adults: A randomized controlled study. *Educational Gerontology, 33*(2), 91–110. doi:10.1080/03601270600846733

Social Security Act. (2018). *Compilation of the social security laws.* Retrieved from http://www.ssa.gov/OP_Home/ssact/comp-ssa.htm

Stout, J. G., Nilanjana, D., Hunsinger, M., & McManus, M. A. (2011). STEMing the tide: Using ingroup experts to inoculate women's self-concept in science, technology, engineering, and mathematics (STEM). *Journal of Personality and Social Psychology, 100*(2), 255–270. doi:10.1037/a002138

Tillinghast, R. C., Petersen, E. A., Fischer, G. L., Sebastian, D., Sadowski, L., & Mansouri, M. (2017). *Expanding STEM outreach through multi-generational reach: Establishing library based STEM programs* Paper presented at the IEEE Integrated STEM Conference (ISEC).

Tsai, H.-y. S., Shillair, R., Cotten, S. R., Winstead, V., & Yost, E. (2015). Getting grandma online: Are tablets the answer for increasing digital inclusion for older adults in the US? *Educational Gerontology, 41*, 695–709. doi:10.1080/03601277.2015.1048165

Turner, P., Turner, S., & Van de Walle, G. (2007). How older people account for their experiences with interactive technology. *Behaviour & Information Technology, 26*(4), 287–296. doi:10.1080/01449290601173499

U.S. Bureau of Labor Statistics. (2018, January 19). *Employment status of the civilian non-institutional population by age, sex, and race*. Retrieved from ftp://ftp.bls.gov/pub/special.requests/lf/aat3.txt

U.S. Equal Employment Opportunity Commission. (1967). *The age discrimination in employment act (ADEA)*. Retrieved from https://www.eeoc.gov/laws/statutes/adea.cfm

Umble, D. Z. (1994). The Amish and the telephone: Resistance and reconstruction. In R. Silverstone & E. Hirsch (Eds.), *Consuming technologies: Media and information in domestic spaces* (pp. 183–194). New York, NY: Routledge. doi:10.4324/9780203401491_chapter_11

United States Social Security Administration. (2018). *Historical background and development of social security*. Retrieved from http://www.ssa.gov/history/briefhistory3.html

van Bronswijk, J., Bouma, H., & Fozard, J. L. (2002). Technology for quality of life: An enriched taxonomy. *Gerontechnology*, *2*(2), 169–172. doi:10.4017/gt.2002.02.02.001.00

van Bronswijk, J., Bouma, H., Fozard, J. L., Kearnes, W., Davison, G. C., & Tuan, P.-C. (2009). Defining gerontechnology for R&D purposes. *Gerontechnology*, *8*(1), 3–10. doi:10.4017/gt.2009.08.01.002.00

van Deursen, A., & Helsper, E. J. (2015). A nuanced understanding of Internet use and non-use among the elderly. *European Journal of Communication*, *30*(2), 171–187.

Van Dijk, J. A. G. M. (2005). *The deepening divide: Inequality in the information society*. Retrieved from https://doi.org/10.4135/9781452229812

van Hoof, J., Kort, H. S. M., Markopolous, P., & Soede, M. (2007). Ambient intelligence, ethics, and privacy. *Gerontechnology*, *6*(3), 155–163. doi:10.4017/gt.2007.06.03.005.00

Van Volkom, M., Stapley, J. C., & Amaturo, V. (2014). Revisiting the digital divide: Generational differences in technology use in everyday life. *North American Journal of Psychology*, *16*(3), 557–574.

Venkatesh, V., Morris, M. G., Davis, G. B., & Davis, F. D. (2003). User acceptance of information technology: Toward a unified view. *MIS Quarterly*, *27*(3), 425–478. doi:10.2307/30036540

Vroman, K., Arthanat, S., & Lysack, C. (2015). "Who over 65 is online?" Older adults' dispositions toward information communication technology. *Computers in Human Behavior*, *43*, 156–166. doi:10.1016/j.chb.2014.10.018

Waldron, V. R., Gitelson, R., & Kelley, D. (2005). Gender differences in social adaptation to a retirement community: Longitudinal changes and the role of mediated communication. *Journal of Applied Gerontology*, *24*(4), 283–298. doi:10.1177/0733464805277122

Weiler, A. (2005). Information-seeking behavior in generation Y students: Motivation, critical thinking, and learning theory. *The Journal of Academic Librarianship*, *31*(1), 46–53. doi:10.1016/j.acalib.2004.09.009

Wengraf, T. (2001). *Qualitative research interviewing: Biographic narrative and semi-structured methods*. Thousand Oaks, CA: Sage.

Wright, D., & Hill, T. (2009). Prescription for trouble: Medicare part D and patterns of computer and internet access among the elderly. *Journal of Aging & Social Policy*, *21*(2), 172–186. doi:10.1080/08959420902732514

Wyatt, S., Henwood, F., Hart, A., & Smith, J. (2005). The digital divide, health information and everyday life. *New Media & Society*, *7*(2), 199–218. doi:10.1177/1461444805050747

Xie, B. (2008). The mutual shaping of online and offline social relationships. *Information Research*, *13*(3).

Xie, B., & Jaeger, P. (2008). Computer training programs for older adults at the public library. *Public Libraries*, *47*(5), 52–59. doi:10.1016/j.lisr.2009.03.004

Yin, R. K. (2009). *Case study research: Design and methods*. Thousand Oaks, CA: Sage.

Yusif, S., & Hafeez-Baig, A. (2016). Older people, assistive technologies, and the barriers to adoption: A systematic review. *International Journal of Medical Informatics*, *94*, 112–116. doi:10.1016/j.ijmedinf.2016.07.004

Zhang, S., Grenhart, W. C. M., McLaughlin, A. C., & Allaire, J. C. (2017). Predicting computer proficiency in older adults. *Computers in Human Behavior*, *67*, 106–112. doi:10.1016/j.chb.2016.11.006

Index